DIRECT CURRENT MOTORS—

Characteristics & Applications

No. 931
$14.95

DIRECT CURRENT MOTORS—
Characteristics & Applications

by Peter Walker

TAB BOOKS
BLUE RIDGE SUMMIT, PA. 17214

FIRST EDITION

FIRST PRINTING—JANUARY 1978
SECOND PRINTING—MARCH 1979

Printed in the United States
of America

Library of Congress Cataloging in Publication Data

Walker, Peter, 1930-
    Direct current motors.

Includes index.
    1. Electric motors, Direct current. I. Title
TK2681.W34   621.46'2   78-13608
ISBN 0-8306-8931-1

Foreword

Direct current motors, like old friends, always seem to be there when you need them. They trace their lineage to the earliest experiments of those profound scholars of the 18th and 19th centuries and today find increasing use with the solid-state revolution. The adage that "The past is prologue" holds true for these machines. Edison's first motors had windings laid directly over smooth, solid-iron cores. Then deep slotted cores with windings placed into them came into use. Today the need for special purpose low-inertia dc motors has led to the development of basket-type armature servomotors, made possible by modern high-strength epoxies. These motors do not use a core. Early motor designers used windings laid *over* a core in a strict application of Ampere's law; today's devotees use the separate-winding technique to minimize mass—when required. Their reason for using the separate winding is for a modern application, but the technique is quite similar. Of course, this is not to say that machines today closely resemble their forerunners. They are as different as a Model T is from the latest Mustang.

These sophisticated advancements are treated in detail throughout the book. The introductory chapter outlines the history of dc machines. It begins with the invention of the compass by the ancient Chinese, when they floated a magnetized needle on a bowl of water. The Greeks used a crude lodestone compass in navigating their sailing ships. In modern times, the trail leads us through the contributions of numerous reasearchers.

Volta's battery made possible the constant source of voltage and current for others to work with. Then in 1819 Hans Christian Oersted, a Danish physicist, discovered that a magentized needle was deflected by a current passing through a nearby wire. Soon afterward, Andre Ampere, a French physicist and mathematician, developed his law through a series of experiments. He discovered that *the force existing between parallel conductors in air when carrying current is directly proportional to their length and the product of their currents and inversely porportional to the distance between them.* In 1831 the English scientist Michael Faraday published in *Philosophical Transactions* the series of "Experimental Researches in Electricity." He had discovered electromagentic induction and built the first dynamo. This work was subjected to rigorous mathematical analysis by the Scottish physicist James Clerk Maxwell, who published the renowned *Electricity and Magnetism.* Later years saw continued work. One of the first to make extensive use of developments in electricity and magnetism was Thomas Alva Edison. He not only produced the first commercially practical incandescent lamp but also developed a complete system of electrical distribution to make the lamp viable. In 1881-82 Edison developed the first central electric-light power plant in the world. Thus it can be seen that practical uses followed theoretical developments very rapidly.

Since the reader, too, must have a thorough understanding of the electromechanical energy conversion process, the fundamental mathematical relationships established by Ampere and Faraday are closely studied early on in the book. With this theory in hand, the book then delves into the practical world of machine construction features. This brings theory and practice together. In addition to considerations of the magnetic circuit, field assembly and brush support mechanisms, there is a complete discussion of construction techniques and testing procedures.

After these general construction features are discussed, the actual classifications of dc machines are presented. Here you will find the old standby series-connected-field motor, the shunt-field motor, the compound-field motor, and the permanent-magnet motor.

Increasingly high costs of energy make efficient use of all machinery mandatory. Thus, anyone involved with the design, specifications, or operation of dc machines needs the provided knowledge of ratings and other pertinent characteristics. Experienced designers know that tests on new motors should closely match theoretical calculations. Probably the most basic rating system for a dc motor is operating voltage, output power, and speed. But many factors can influence the operation of the motor. One of the most obvious is temperature. Most motors are rated for an ambient temperature of 40°C. Operation at ambient temperatures much above this must be derated. In fact, if speed is a critical specification, a motor may have to be designed for exact speed at a given operating temperature. This whole question of temperature and motor operation is so important that the author devotes considerable space to an analysis of it. Closely related are cooling and machine enclosures. Many configurations are available, but deciding upon the best one for maximum efficiency at reasonable cost can be a problem. All the various classifications are discussed, and the pros and cons of each are spelled out.

Increasing use of dc motors with solid-state devices means that machine characteristics of little interest years ago now take on importance. Impedance of the windings must be known before maximum efficiency can be realized from solid-state controls. In fact, deciding whether a machine will operate properly under various conditions requires the use of proven analytical methods using equivalent circuits. A number of these methods are discussed—including those for the popular shunt-field motor and the permanent-magnet motor. Of particular interest in these evaluations are the effects of brushes on motor performance.

This aspect of motor performance is of such importance that a whole chapter is devoted to the subject of commutation, sparking, and brush wear. The problem of excessive brush sparking appeared with the introduction of deep-slotted armatures about 1890. Those studying the problem offer differing views of its exact nature. The presentation of these differing viewpoints allows the reader to acquire an overall understanding of the subject that would not be possible if the author had concentrated on a single aspect.

Whether the book serves as a primer on dc machinces or as a reference to resolve problems, the reader will find that this book presents an invaluable background in dc motors.

TAB Editorial Staff

Contents

Chapter 1

History and Development of the DC Machine

In our modern day world, electric machinery takes on many shapes and sizes. The many highly specialized forms which have evolved for particular application requirements oftentimes seem to suggest, at least to the man on the street, no common roots of origin. But despite the varied forms of modern machinery design, there is one type of motor, the dc motor, which retains a simplicity of concept that easily allows its evolution to be traced back to the primitive machines of Barlow and Faraday in the early 19th century. This fact is ironic because as will be seen in succeeding chapters, the dc motor, while it does retain a simplicity of concept, will also assume the most radical of configurations and allow usage with a greater variety of power sources and controllers than any other motor type.

The reason for the continued broad usage and design evolution of the dc motor is quite simple. Its unique characteristics—a speed proportional to applied voltage, increasing torque with decreasing speed, plus the ability to run off storage batteries—make it a logical choice for many electric drive applications. Coupled with the unique torque-speed characteristics of the dc motor are a number of other factors which have increased the level of interest in recent years. One factor is the technological breakthrough which has occurred in the area of power electronics, including SCRs, power transistors, and integrated circuits. These solid state electronic devices have allowed the development of many new motor controllers with improved performance and lowered costs. Still another factor in the renewed dc motor interest has been the development of improved motor features through the use of newly developed materials. In this category fall the recent proliferation of permanent-magnet motors made possible by the development of barium-ferrite magnets and the high performance, low-inertia basket type armature servomotors made possible by high strength epoxies.

To fully appreciate and understand the variegated forms of the modern dc motor, it is useful and interesting to consider its development from a historical perspective.

EARLY DISCOVERIES AND HISTORY

Invention of the electric motor was anteceded by man's discovery of magnetic and electric phenomena. Knowledge of these phenomena was very slow in developing and it was not until the 19th century that the two effects were shown to be related. Knowledge of and use of permanent magnets can be traced back into ancient times. The ancient Chinese are known to have invented a compass by floating a magnetized needle in a bowl of water. Similarly, the ancient Greeks used a crude lodestone compass as an aid to navigation on their sailing ships. However, it took many more hundreds of years before man discovered the vital connection between electricity and magnetism.

The knowledge of magnetism and electricity was advanced greatly during the late 16th and early 17th centuries. During this period of time the famous English scientist Sir William Gilbert made many experiments with static electricity and magnets but did not connect the two phenomena. Also during this period of time it was discovered that electricity could be made to flow over substantial distances through "conductor" materials. The nature of magnetism and electricity was slowly being discovered and documented. At that time, however, the materials and energy sources available to work with had severe limitations. In particular there was not a source of electricity capable of sustaining a current flow for any appreciable period of time. This limitation did, of course, severely limit the experimenting that it was possible to conduct.

The next major breakthrough of technology was the invention of the storage cell by Volta, around 1790. This invention was based upon a long series of discoveries by Galvani, Volta, and others who believed chemical action could provide a means for storing and retrieving electricity. The cell provided by Volta, while a far cry from today's batteries of tremendous ampere-hour capacity, did nevertheless give the scientist of his day a source of electricity which could sustain a current long enough to make meaningful experiments. The consequences of Volta's invention were dramatic and relatively quick to follow. It made possible the discoveries of electrodynamics which in turn were quickly applied to the invention of an electric motor, a process of development which continues today.

THE NATURE OF ELECTROMAGENTISM

The Danish physicist Oersted is credited with making the discovery of a relationship between magnetism and electricity. Oersted's experiment (Fig. 1-1) involved bringing a current-carrying conductor close to a magnetized needle. When the conductor was held parallel to the needle, the needle was observed to deflect. When the polarity of the current was reversed, it was seen that the needle again deflected, but in the opposite direction. From this simple experiment it was possible to deduce that the electric current in some way was responsible for applying a force to the needle. It was a simple deduction with great significance.

In 1820, Oersted published his findings in a paper which was presented to the French Academy of Science. In attendance during the presentation of the Oersted paper was a French mathematician by the name of Andre Marie Ampere. The concept suggested to

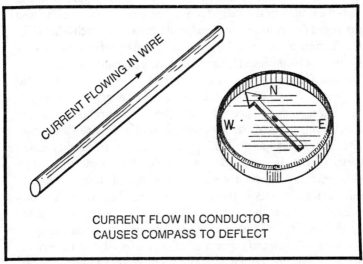

CURRENT FLOW IN CONDUCTOR
CAUSES COMPASS TO DEFLECT

Fig. 1-1. Oersted's discovery that electric current causes a compass to deflect.

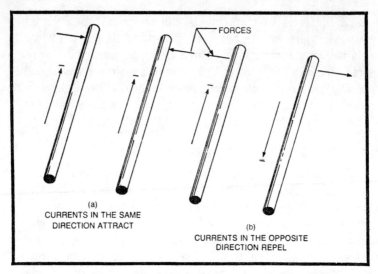

Fig. 1-2. Ampere's law. Current carrying conductors exert force upon one another.

Ampere by Oersted's discovery, a force acting through a distance with no visible means of transfer, so intrigued him that he left the presentation and took up a similar line of experimentation. Within a week's time, Ampere has not only confirmed Oersted's results but also significantly advanced the knowledge. He found that the force exerted on a conductor was proportional to the strength of the magnet and also to the magnitude of the electric current. He also discovered that two currents produced a force upon each other (Fig. 1-2). Ampere discovered the attractive force between two currents as a result of his powerful logic and deduction. It had been shown that a current and a magnet developed a force relationship. It had also been known for a long time that two magnets exerted a force upon each other. Ampere speculated whether two currents would also exhibit a similar force phenomenon when brought into close proximity. To test this speculation Ampere quickly devised and executed an experiment. He suspended two parallel wire conductors with a slight separation between them. He found that when the conductors both carried currents in the same direction, they moved towards each other. When the current in one of the conductors was reversed, the wires moved apart. Over the next fifteen years Ampere continued his work and experimentation. He found that if a current-carrying conductor were wound into a long coil of smaller diameter, it would exactly duplicate the effects of a long bar magnet. Ampere even speculated that a magent's properites were due to current flow within the magnet material. A speculation that is in fact consistent with the modern electron spin theory of magnetism. The culmination of Ampere's work was a treatise on electrodynamics wherein he related in mathematical form electric and magentic force phenomena. These relationships continue to provide the basis for the teaching of electrodynamic principles in our modern day world.

The work of Ampere definitely established that mechanical forces could be developed predictably by a suitable arrangement of current carrying conductors in the vicinity of a magnet. Ampere's work served to point the way for a group of practical experimenters who were quick to show the potential of the discovery in many applications. Among the most important of these, as far as motor development is concerned, were Barlow and Faraday. The Barlow wheel (Fig. 1-3) was invented about 1822 and can be regarded as the first successful attempt by man to utilize the newly discovered

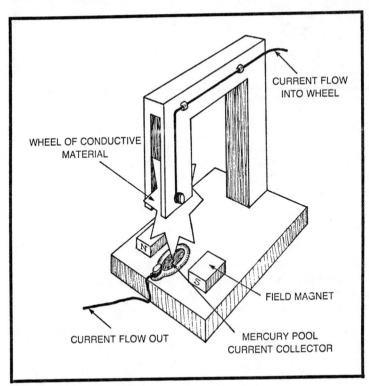

CURRENT FLOW INTO WHEEL

WHEEL OF CONDUCTIVE MATERIAL

N

S

FIELD MAGNET

CURRENT FLOW OUT

MERCURY POOL CURRENT COLLECTOR

Fig. 1-3. The Barlow wheel. An early dc machine design.

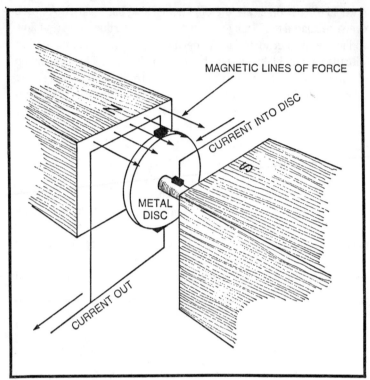

Fig. 1-4. The Faraday disc. The first dc generator.

electromagnetic forces to create an electric motor capable of continuous rotation. The invention of Barlow's was followed in 1831 by
the invention by Faraday of his disc machine (Fig. 1-4). The Faraday
machine was, in its operating principles, an improvement on the
Barlow machine. The Faraday machine consisted of a disc made of a
conductor material mounted to a shaft and mounted in a magnetic
field. Faraday found that when the disc was cranked he could, by
means of sliding contacts, measure a voltage from the center to an
outer edge of the disc. Faraday's machine was, in fact, the first
electric generator. The concept of Faraday's machine is preserved
in modern homopolar, or unipolar, machine.

Faraday also contributed to the knowledge of electricity and
magnetism by recognizing that the magnitude of voltage produced by
his machine was proportional to the speed at which it was turned. He
also found that changing the strength of the magnet would change
the generated voltage in the same manner. Faraday is thus given
credit for formulating the laws of electromagnetic induction. Fara-

day, however, was not a formally trained scientist and was not able to develop a rigorous theory for his machine. It took succeeding theoreticians to fully develop a theory to explain the effects observed.

FORCE AT A DISTANCE AND ELECTROMAGNETIC THEORY

An interesting sidelight to the history of electric motor development is the different viewpoints of electromagnetic force that were held by the scientists of Faraday's time. For although their knowledge of electric and magnetic phenomena was far from complete, these early experimenters included some of the best intellects the world has ever known. In fact a fairly common characteristic of the early scientist was his dual role as a philosopher. Thus as a new phenomenon was discovered, the scientist of the day made a great effort to fully understand it and relate it to the world around him.

So it was almost a certainty (with the intellectual curiosity of the scientist of the time) that the forces defined by Oersted and Ampere would become a topic of philosophic argument. The big question was, by what medium was the force transmitted to the point of measurement? After contemplating this question for a long time, Ampere decided that the best approach was not to attempt to describe the medium for force transmittal. Rather, Ampere concentrated on properly identifying and relating all the factors involved in electromagnetics. Ampere's viewpoint came to be known as the "Direct Action at a Distance Theory of Electrodynamics." This point of view is largely that of the mathematician whose main concern was the development of equations elegant enough to explain the observations.

Although Faraday's work was based upon the principles of Ampere, he could not accept the "Action at a Distance Theory." To his practical mind, to feel a force meant that something has to be exerting the force. In order to fill the void in the prevailing theory, Faraday proposed that there must exist throughout all of space a substance which could be neither seen, nor felt, nor weighed. This substance came to be known as the ether. If such a substance did exist, Faraday proposed a model which could account for the transmittal of forces. In Faraday's model the ether became the medium for the establishment of lines of force. In the space around a magnet, the lines of force extended from one pole of the magnet to the other. These lines of force came under a condition of stress due to the magnetic influence and thus could be regarded as a form of stored mechanical energy. The nature of the energy was quite similar to that possessed by a coiled spring. The lines of force were pictured as having certain mechanical properties. They were capable of being stressed and deformed by the presence of iron. Since an electric current was known to have some of the properties of a magnet, lines of force would also exist in the space around the current. Mechanical force could then be regarded as the result of an interaction of the two force fields. While Faraday's model was the product of his logical deductions, he was not able, because of his lack of expertise with mathematics, to develop equations describing the model that would satisfy the theoreticians of his day.

The development of a theory to describe Faraday's model was left to James Clerk Maxwell. Maxwell was born in 1831 (the same year Faraday invented his machine) in Scotland. During his undergraduate days, Maxwell read the experiments that Faraday had conducted and also studied the writings of Ampere. After graduating from college with a doctorate in physics, Maxwell accepted a professorship at the University of London. At that time Faraday was still active as a researcher in London and the two men met and became friends. There is no doubt that Faraday was able to influence the thought of Maxwell. The result was the publication in 1864 by Maxwell of his "Treatise on Electricity and Magnetism." This work is still regarded today as the definitive work on electric and magnetic field phenomena. Maxwell's theory, which is commonly known as field theory, reduced Faraday's concept to mathematical language by use of vector calculus. In its application to electric currents, Maxwell's theory argued that each current established a field in the space about it through an unknown medium (the ether). The force fields due to different currents would in turn interact and produce the resulting mechanical force that could be observed and measured. The force field due to an electric current was similar in nature to that produced by a magnet. The interaction of the fields of magnet and current thus explained the forces discovered by Oersted and Ampere.

Maxwell's theory was so complete and general in nature that it applied equally well to steady (dc) fields or to time varying (ac) fields. Maxwell's field theory suggested the possibility of energy radiation by high frequency ac fields more than 20 years prior to the discovery by Hertz of radio waves in 1887. The work of Maxwell strongly suggested but did not prove the existence of an ether. During the years following the publication of the treatise, many elaborate experiments were devised and executed to detect the presence of the ether but all to no avail. The quest for the ether was in turn taken up at the end of the 19th century by a young physicist named Albert Einstein who attempted to extend the work. In 1905, Einstein published his classic paper "On the Electrodynamics of Moving Bodies," in which he attempted to explain the various experimental results obtained by others in their search for the ether. Einstein also failed to prove the existence of an ether in his paper, although it did otherwise support the field theory.

COMMERCIAL DEVELOPMENT OF ELECTRIC MACHINES

While the controversy over the most appropriate theory raged among theoreticians in the academic community, an increasing number of practical men were applying the newly discovered principles to the making of a commercial product. The result was that a new industry was developing. In 1880, Tom Edison started the large scale production of dc machinery at the Edison Machine Works in New York City. A number of other companies were started that same year and by 1887 there were 15 companies in the United States involved in the manufacture of electric machines. These pioneer manufacturers had an annual output of 10,000 units in sizes up to 15 hp.

These early machines were very unlike the motors of today. The machines were designed with the windings laid directly onto the surface of a smooth, solid iron cyclinder. The conductors were secured firmly with a cord and depended upon friciton between the conductors and the cylinder to transmit torque to the output shaft. This design worked well enough on smaller motors (and has, in fact, been resurrected with modern high performance servomotors) but as motors got larger and forces greater, the conductors often broke loose from the rotor surfaces with a resulting disaster. An improved design was then put into production which utilized wooden pegs (drive pins) set into the rotor surface to restrain the conductors and provide a surer means of transmitting force to the shaft. The drive pin design (Fig. 1-5) was effective and allowed larger machines to be built.

It should be kept in mind that in building the motors with conductors mounted to a smooth surface, the engineers of the day were strictly adhering to Ampere's theory of electrodynamics. The theory, as they understood, required the current-carrying conductor to take up the force and to be located in the area of high

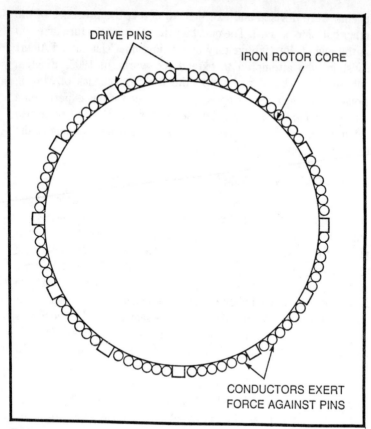

DRIVE PINS

IRON ROTOR CORE

CONDUCTORS EXERT
FORCE AGAINST PINS

Fig. 1-5. Drive-pin design used to transmit torque.

magnetism between the iron members of the motor (Fig. 1-6). Up to this time the theory had been adequate to allow machines to be designed with a reasonable confidence in the result and the theory was well established. By 1890, motors were being built with an iron cylinder composed of laminations of sheet steel instead of solid iron. From this point the next step was to incorporate the drive-pin function into the steel lamination by punching slots into the periphery (Fig. 1-7). To the motor engineer of this era the use of a deep slot in a steel structure suggested a reduced electromagnetic force. This conclusion seemed to follow by a strict adherence to Ampere's law. For if steel or iron were put around the current, it would act to divert the magnetism from the space occupied by the current. The result should be that the current in a space of less magnetism would experience a lesser force. To the surprise of most motor engineers,

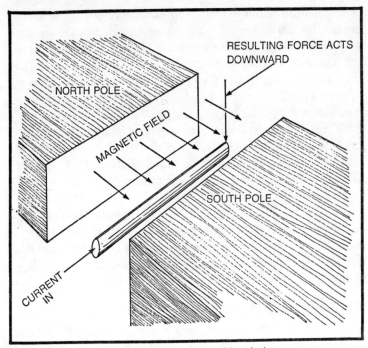

Fig. 1-6. Ampere's law as it applies to dc machine design.

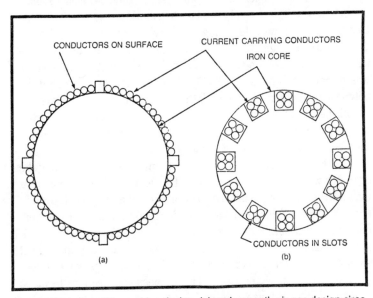

Fig. 1-7. Evolution of dc machine design, (a) early smooth air-gap design circa 1870 and (b) deep slot design of today's machine.

the first motors built with slotted cores were tested and found to develop the same force as if the current existed in the air gap between two smooth steel structures. Special test motors were built which allowed the measurement of the actual force on the conductors in a slot. It was found that the conductors themselves (when in a slot) experienced just a small part of the total force comprising the machine output.

The enigma of the slotted core continued for a number of years until 1896. In that year Dr. G. F. Searle dusted off the theory of Faraday and Maxwell and came up with a solution. Searle showed that when Maxwell's field theory was properly applied, the calculated electromagnet forces would agree with measured values even if the conductors are embedded in the iron. He further showed that the mechanical forces were actually experienced in those regions where the lines of force existed, in this case the iron around the slot. This problem solution was clearly a vindication of the Faraday-Maxwell concept.

THE FALL AND RISE OF THE DC MACHINES

Up until the 1890s, the great majority of electric machines were direct current. This situation was changed rather quickly, however, by several technological developments. In 1886, William Stanley demonstrated the versatility of an ac distribution system utilizing the transformers he was designing. Stanley's demonstration showed that electric energy was most conveniently transmitted and distributed in an alternating current form. The shape of things to come was decided with finality in 1888, when Nikola Tesla invented the polyphase induction machine. With the advantage of an ac system established and the availability of an electric motor to use ac electric power, it took just a short time to convert the earlier dc system to the 60 hertz system we know today.

With the change to an ac power system it was inevitable that the dc motor would also fall into decline. The ac motor, with its inherent manufacturing cost advantage, has proliferated in some form into every facet of modern day life. The dc motor continued to be built during the first half of the 20th century but as a decreasing percentage of total electric motor production. The continued production of dc motors was required for certain special applications where it was the only type that would work. These special applications included those where only battery electric supply was available, or where closely regulated speed control was required.

22

During the past 15 to 20 years there has been a revival of interest in dc machines. The reasons for this renewed interest have already been alluded to. They include the demands for automatic production equipment to increase the productivity of industry. The demands of automation again place a premium on the unique controllability characteristics of the dc motor. They also are due to technological developments in other fields (such as power electronics) which have allowed the development of new controller designs which in turn have produced new types of equipment. Finally, in the last five years there have been powerful forces, economic, regulatory, and political, which are acting to create a favorable atmosphere for a rapidly growing usage of dc machines. Events such as a shortage of gasoline and a growing awareness of the finite limitations of our petroleum resources will certainly push R & D efforts in the direction of an electric vehicle to replace present gasoline powered automobiles. It is expected that the dc machine in its various specialized forms will be increasingly important in the years ahead.

Chapter 2

Electromechanical Energy Conversion

The process of energy conversion in an electric machine is based upon the natural effects investigated by Ampere and Faraday. An understanding of the conversion process may be obtained by considering the fundamental mathematical relationships established by Faraday and Ampere. These simple equations in elemental form can then be operated upon and from them working formulae derived which will be applicable to practical machines. These derived expressions for voltage and torque are the tools by which the engineer starts a design procedure or analyzes a machine performance. Table 2-1 defines the symbols used in this chapter.

THE VOLTAGE EQUATION

In its most basic form, Faraday's law of Electromagnetic Induction is given by the expression,

$$e = \frac{d\Phi}{dt} \tag{2.1}$$

This expression states that the instantaneous value of a generated (or induced) voltage in a circuit consisting of a single turn of wire is equal to the rate at which the magnetic flux enclosed by the circuit is changing with respect to time. A correct inference is that if Eq. 2.1 gives the voltage of a single-turn circuit, the voltage will be increased if the number of turns are increased. If the single turn circuit

Table 2-1. Chapter 2 Symbology

a—the number of current paths through the armature circuit
B—the magnetic flux density given as units of magnetic flux per unit area
e—instantaneous generated voltage, may be a time varying value
E—average or dc generated voltage
f, F—electromagnetic force developed on a current carrying conductor in a magnetic field with units of newtons or pounds
i, I—electric current in units of amperes
l—length of a current carrying conductor in units of meters or inches
N—number of turns in electric coil
n—motor rotational speed in revolutions per minute
P—number of magnet poles in a dc machine
R—electrical resistance of a winding assemblage in units of ohms
r—the effective radius at which a rotating coil is located in units of meters or inches
T—electromagnetic torque in units of meter-newtons or foot-pounds
Z—the number of active conductors in an armature circuit
η—the ratio of output power to input power expressed as a percentage of the input and defined as the efficiency of a machine
ψ—the mechanical radians of one pole pitch in an electric machine
Φ—magnetic flux per pole may be in units of webers or lines
θ—an angular increment around the periphery of a machine's air gap in radians

represented by Eq. 2.1 is replaced by a coil of N turns, Eq. 2.1 can be written as

$$e = N \ \frac{d\Phi}{dt} \tag{2.2}$$

Equation 2.2 is a general equation where the voltage may be developed by different means. It could be similar to that induced in a stationary transformer coil when ac voltages are used to provide a magnetic field which varies with time. Or it will equally well describe the voltage generated in a motor winding which is moving relative to a constant magnetic field. It is this second type of voltage generation that is of special interest to this discussion. It is necessary then to put Eq. 2.2 into a form which will introduce the rotational speed characteristic along with the other parameters associated with a machine. This is done as follows:

$$\frac{d\Phi}{dt} = \frac{d\Phi}{\Omega} \cdot \frac{d\theta}{dt} \tag{2.3}$$

and

$$\frac{d\theta}{dt} = \frac{P}{2} \ \frac{2\pi n}{60} \tag{2.4}$$

The identities of Eqs. 2.3 and 2.4 are then substituted into Eq. 2.2 to obtain

$$e = N \ \frac{P}{2} \ \frac{2\pi n}{60} \ \frac{d\Phi}{d\theta} \tag{2.5}$$

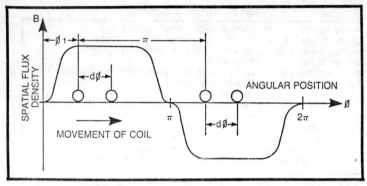

Fig. 2-1. Distribution of magnetic flux density and conductors in the air gap of a dc machine.

A suitable expression for $d\Phi$, relating it to machine parameters, can be obtained by considering Fig. 2-1. The two sides of the coil are identified respectively as c and c^1. The coil sides are shown as they would move if they were affixed to a rotor in an actual machine. The movement is relative to a field of constant strength which is fixed in space. It can be seen that as the coil sides are moved by a small angle $(d\Omega)$, the change in flux with time will be

$$d\Phi = 2B\,d\theta \qquad (2.6)$$

The expression for $d\Phi$ can be substituted into Eq. 2.5 to produce

$$e\,d\theta = N\left(\frac{P}{2}\right)\left(\frac{2\pi n}{60}\right)\ \left(2B\,d\theta\right) \qquad (2.7)$$

In order to find the dc, or average, generated voltage, Eq. 2.7 is integrated over the interval of 0 to π radians and the integral is then divided by π. The result is,

$$E = \frac{1}{\pi}\int_0^\pi e\,d\theta = \frac{1}{\pi}\int_0^\pi N\left(\frac{P}{2}\right)\left(\frac{2\pi n}{60}\right)\quad 2B\,d\theta$$

$$E = \frac{2NPn}{60}\int_0^\pi B\,d\theta \qquad (2.8)$$

It is recognized by again referring to Fig. 2-1 that the flux per pole, Φ, is given by

$$\Phi = \int_0^\pi B\,d\theta$$

so that Eq. 2.8 becomes

$$E = \frac{2NPn\Phi}{60} \qquad (2.9)$$

Equation 2.9 allows calculation of the voltage generated in a coil of N turns of wire when it is rotated at n revolutions/minute in a machine of P magnetic poles with Φ amount of flux per pole. Of course, a practical machine will usually have a number of such coils connected in some manner. Since the calculated coil voltage is an average value, it follows that the additional coil voltages will add arithmetically if they are connected in series. The voltage equation for a complete armature winding is then written as

$$E = \frac{PZ\Phi n}{60a} \quad \text{volts} \qquad (2.10)$$

or

$$E = \frac{PZ\Phi n}{60a} \times 10^{-8} \qquad (2.10a)$$

where Z is the total conductors in the armature winding and a is the number of current paths in the circuit. The polarity of the generated voltage is always such as to produce a current that develops a force to oppose the movement.

Figure 2-2 shows a conductor moving parallel to the face of the north pole of a magnet. As the conductor cuts the magnetic lines of force a voltage is induced. An easy way to determine the polarity of the induced voltage is as follows. Consider the conductor to be a right-handed screw. Turn the arrow representing the direction of conductor movement until it coincides with the direction of magnetic flux. The polarity of the induced voltage will act in the same direction as the screw advances. Thus, for the directions shown in Fig. 2-2, the screw is turned in a counterclockwise rotation and there would be a resulting movement from left to right. The induced voltage also acts from left to right.

THE TORQUE EQUATION

An expression for the torque developed in an electric machine can be derived directly from Ampere's law of electromagnetic force. Ampere stated that the force exerted upon a current carrying conductor in a magnetic field was proportional to the strength of the

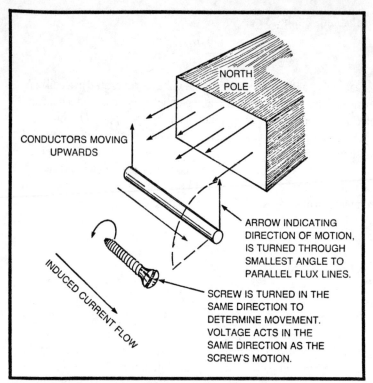

NORTH POLE

CONDUCTORS MOVING UPWARDS

ARROW INDICATING DIRECTION OF MOTION, IS TURNED THROUGH SMALLEST ANGLE TO PARALLEL FLUX LINES.

SCREW IS TURNED IN THE SAME DIRECTION TO DETERMINE MOVEMENT. VOLTAGE ACTS IN THE SAME DIRECTION AS THE SCREW'S MOTION.

INDUCED CURRENT FLOW

Fig. 2-2. Rule for determining the polarity of generated voltages.

magnetic field, the magnitude of the current, and the length of the conductor. This relationship can be expressed as

$$f = Bli \qquad (2.11)$$

The torque associated with a motor will be given by the electromagnetic force multiplied by the radius of the rotor. Equation 2.11 then becomes

$$T = rf = rBli \qquad (2.12)$$

Now it is recognized that the flux per pole is given by

$$\Phi = Blr$$

and Ψ which is the angle in radians in one magnetic pole in given by

$$\Psi = \frac{2\pi}{P}$$

If the appropriate substitutions are made in Eq. 2.12 a new expression is obtained

$$T = r \left(\frac{\Phi P}{2\pi r} \right) i$$

or

$$T = \frac{P\Phi i}{2\pi} \qquad (2.13)$$

Equation 2.13 gives the torque developed by a single conductor. The expression can be modified so that it can be applied to practical machines which have many conductors in an armature winding by multiplying Eq. 2.13 by the number of conductors Z. Additional attention must be given to the fact that i is the conductor current. In order to arrive at a formula with the total current as a factor, it is recognized that total armature current will be given by the conductor current multiplied by the number of current paths through the armature.

Or

$$I = ia$$

If these additional modifications are performed on Eq. 2.13 a final form is obtained which is

$$T = \frac{PZ\Phi I}{2\pi a} \quad \text{[meter-newtons]} \qquad (2.14)$$

An alternate form of Eq. 2.14 can be written in terms of the more commonly used English units of torque, pound-feet, by using the proper conversion factors. Then Eq. 2.14 can be rewritten as

$$T = \frac{0.117 \, PZ\Phi I \times 10^{-8}}{a} \quad \text{[pound-feet]} \qquad (2.15)$$

(In Eq. 2.15 the flux Φ per pole is given in units of maxwells or lines which requires the factor 10^{-8}.)

The direction of the torque (or force) so developed is determined by the direction of current flow and the polarity of the magnetic field. These parameters have a relationship that can be indicated vectorially by the equation

$$\mathbf{F} = \mathbf{I} \times \mathbf{B} \qquad (2.16)$$

29

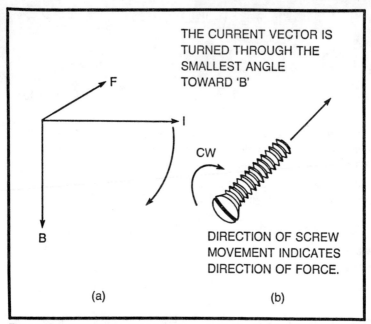

THE CURRENT VECTOR IS
TURNED THROUGH THE
SMALLEST ANGLE
TOWARD 'B'

F

I

CW

DIRECTION OF SCREW
MOVEMENT INDICATES
DIRECTION OF FORCE.

B

(a)   (b)

Fig. 2-3. Orientation of developed electromagnetic forces, (a) vector representation of magnetic field, current, and force; and (b) the right-hand screw analogy.

The determination of the direction of the developed torque (force) can also be helped by a right-hand screw analogy. Look at Fig. 2-3. Figure 2-3(a) shows the vector representation for Eq. 2.16. If current vector **I** is turned to coincide with field vector **B** it will move in a clockwise rotation. The clockwise rotation will cause the screw shown in Fig. 2-3(B) to move forward in the direction as shown by the force vector **F**.

MOTOR AND GENERATOR ACTION

From preceding discussions of electromagnetic induction it may already have been surmised that the generation of a voltage and development of a torque are interrelated. Such is indeed the case. Whenever a conductor is moved through a magnetic field a voltage is generated along the length of the conductor. This is true if the conductor happens to be a part of either a motor or a generator. Similarly, whenever a conductor carries an electric current while in a magnetic field, mechanical forces are developed. Again, this effect is true regardless of whether the conductor is part of a generator or a motor. These two effects are the essence of electromechanical energy conversion and are related.

30

If a generator is connected to some electrical load so that current flows through the armature windings, the conditions exist for the development of a torque. In this case the developed torque is a negative one which opposes the rotation of the prime mover. If the electrical loading is increased so as to increase armature current, the negative torque will also grow proportionately. The result is that the prime mover engine will be slowed and possibly even stalled out.

An analogous effect is observed when electric motors are run. When a voltage is applied to a motor at standstill there is no voltage being generated and the resulting armature current will be very high. The current-carrying conductors in a magnetic field develop torque, and since the current is very high, the resulting torque is also high. The torque causes the motor to accelerate to some rotational speed. As the motor comes up to speed, the conductors which are now moving through a magnetic field will experience an induced voltage which increases with speed. The induced voltage has a polarity which opposes the applied voltage. The generated voltage acts to reduce current flow with resulting reduction of torque development. The final speed of a motor will then be the one which establishes an equilibrium condition between the factors of applied voltage, generated voltage, and load torque.

If generator and motor effects occur simultaneously in a machine, the question naturally occurs. "Can a machine be used as either a motor or a generator as the situation may require?" The answer is yes. Energy flow in an electric machine is completely reversible. Refer to Fig. 2-4. The figure illustrates that if mechanical energy is put into a machine, generator action will occur and electric energy is obtained as an output. On the other hand, if electric energy is put into a machine, motor action will occur and mechanical energy output is available. In practical machines, which are usually designed for a specific purpose, there will often be some design differences between machines designed as motors and those designed as generators. The reason is to enhance particular characteristics that are desired. The principles of operation, voltage and torque generation, however, are the same for either type of machine.

COMMUTATION IN A DC MACHINE

All dc machines are characterized by some form of commutation, or switching, function. The need for the commutation function can be understood by consideration of the principles of force development in a magnetic field. In the discussion of forces previously,

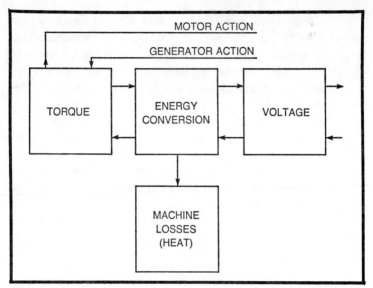

Fig. 2-4. Energy flow in an electric machine.

it was brought out that the direction of force was determined by the directions of the magnetic field and electric current. If either field or current were to change there would be a corresponding change in the direction of the developed force. Obviously, to be a useful device, a motor must develop a steady torque acting in one direction of rotation.

To provide this condition of steady torque development it is necessary that the direction of current flow remain constant with respect to the magnetic field. This arrangement of current and magnetic field is shown in Fig. 2-5 for a two pole motor. The convention used shows current flowing into the paper in the conductors over the top half of the rotating member. Similarly, current flows out of the conductors which are located around the bottom half. This figure illustrates an important principle of dc motor design. It is necessary that different current directions exist opposite poles of different magnetic polarity. With the orientation as shown in this figure a constant torque is developed for ccw rotation. If the rotor were allowed to turn 90°, it would then be as shown in Fig. 2-6. In this position there would be a zero net output torque because of the opposite current directions under the same magnet pole. These conductors develop forces that cancel each other out.

If the rotor were turned another 90° a condition would exist where current directions were all the same again, opposite a particular magnet pole. This condition is shown in Fig. 2-7. In this situation, the currents are exactly opposite of Fig. 2-5, while the magnetic field remains the same. The result is that in Fig. 2-7, torque is developed that would cause a clockwise rotation.

From the foregoing it is clear that a dc motor needs a device that will switch the direction of current flow in its conductors as they rotate past the influence of one magnetic pole and into the influence of one of opposite polarity. It can also be deduced from an inspection of the Figs. 2-5, 2-6, and 2-7, that the proper angular position for the switching, or commutation, to occur is at the point halfway between the north pole and the south pole.

In a conventional dc motor this switching function is provided by the use of sliding contacts. These sliding contacts consist of a set of stationary brushes and a rotating current collector, or as it is com-

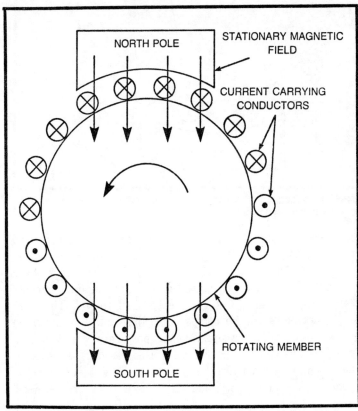

Fig. 2-5. Torque development in a dc motor. Torque developed in counter-clockwise direction.

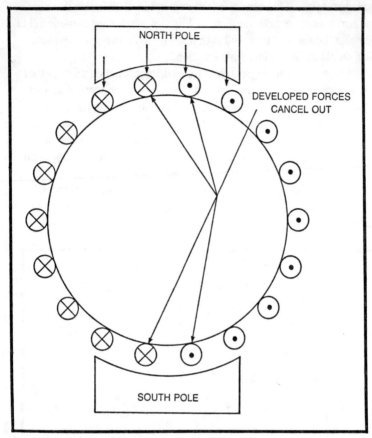

Fig. 2-6. Disposition of current carrying conductors with the brush axis made to coincide with the main field axis.

monly called, a commutator. The commutator most commonly used is a cylinder composed of segments of conductive material (copper) which are insulated from one another (Fig. 2-8). The ends of the armature coils are then connected to the commutator segments according to how their voltages are required to add. The commutator segments can be compared to the contact points of an electrical switch. In this case the contact points are perceived to be in continuous movement, making and breaking contact, as the motor rotates. This repeated interruption of current flow causes a number of problems for the dc motor and will be discussed in greater detail later.

The function of the commutator and brushes can be more clearly understood by looking at Fig. 2-9(a). The small arrows indicate the direction of current flow through the individual coils of the armature. It is seen that current flows in at brush A and follows two parallel currents through the armature coils and finally out at brush B. The one path is provided by coils 1-1', 2-2' and 3-3'. The second path is provided by coils 6'-6, 5'-5, and 4'-4. It is noted that the current flow is from 1 to 1' in coil 1. Now if the armature is rotated ccw by a small angle, the situation is as shown in Fig. 2-9(b). In this position coils 1 and 6 are seen to be shorted by the brushes that contact segments 1 and 2 for brush A and segments 4 and 5 for brush B. In this position the current in the short circuited coils is

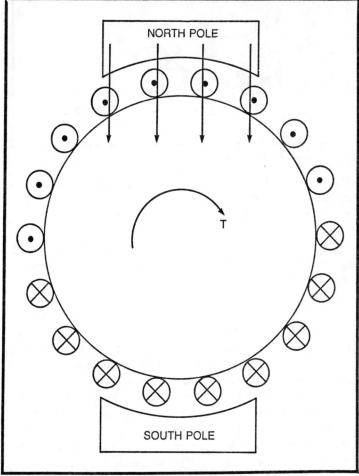

Fig. 2-7. Torque development with brush axis turned 180° from that of Fig. 2-5. Direction of torque in clockwise direction.

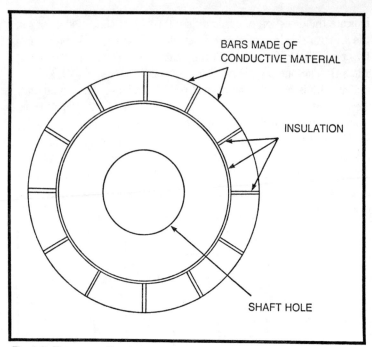

Fig. 2-8. The conventional commutator construction.

being commutated. The coil current will begin to fall towards zero from the value it had in the previous position of Fig. 2-9(a). The current through the remainder of the armature circuit remains unchanged.

Figure 2-9(c) shows the armature rotated by an additional small angular increment. In this position brush A contacts only segment 2, and brush B contacts only segment 5. The current in coil 1 is now fully commutated. Close inspection shows that current flow is now from 1' to 1 in coil 1, just the opposite of what it was in Fig. 2-9(a). The same reversal of current flow has occurred in the coil which has passed under brush B, which is coil 4 for the positions illustrated.

The sequence of illustrations in Fig. 2-9 describes the function of commutation. As a consequence of reversal of current flow as the commutator segments pass under the brush, the armature current has a fixed position in space. The current distribution of Fig. 2-5 is then seen to be achieved if the brush axis is positioned at right angles to the main field axis. In this position all conductors contribute additive electromagnetic forces. This position is thus seen to be the optimum for development of maximum torque.

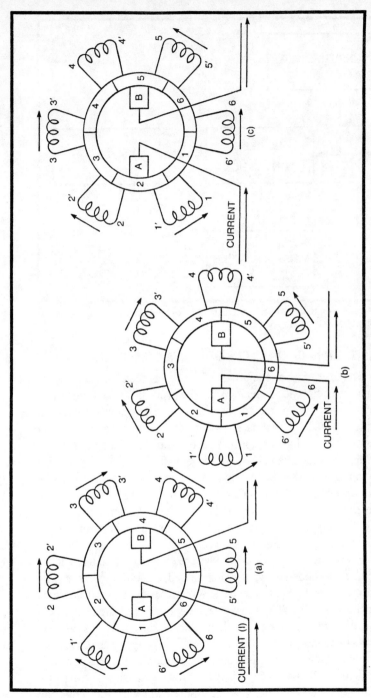

Fig. 2-9. The function of the commutator in changing current-flow direction (a), (b), (c).

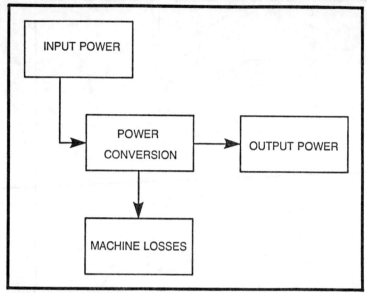

Fig. 2-10. Power balance in an electric machine. The input is equal to the sum of the output plus the machine losses.

CONSERVATION OF ENERGY IN AN ELECTRIC MACHINE

The preceding sections described the process of energy conversion in an electric machine. While the process may at times seem a mysterious one (refer to the Force at a Distance...section of Chapter 1) it does fortunately follow another fundamental law of physics, viz, the Law of Conservation of Energy. This consistency with another fundamental law does in fact afford an alternate basis of developing analytical expressions for motor characteristics. However, since the equations for torque and voltage development which have previously been developed from electrodynamic theory are perfectly adequate, it will suffice here to show how an energy balance does apply to an electric machine.

Every electric machine is characterized by an input and an output. As has been previously mentioned, the energy flow may be from mechanical form to electrical (generator action) or vice versa (motor action). It will be more convenient and meaningful to talk in terms of units of power rather than energy. Since power is the time derivative of energy, it will be perfectly valid to apply the principle of conservation of energy to an analysis of conservation of power.

Figure 2-10 shows pictorially the power balance which exists in a dc machine. It is obvious that the input power must equal the sum of output power plus the machine losses. The ratio of output power to the input power is a commonly applied figure of merit used to express the efficiency of the machine. Then if units of watts are used to express power, the machine efficiency will be given by

$$\eta = \frac{\text{watts out}}{\text{watts in}} \times 100\% \qquad (2.16)$$

The difference between the input watts and the output watts represents the losses internal to the machine. The efficiency of a machine, or even more directly the losses in a machine, is critical in determining a machine's rating. This is because all of the machine losses will ultimately be dissipated as heat. The amount of heat generated will, in turn, determine, for a given motor frame and ambient condition, how hot the motor gets. Since a practical machine is limited to some definite maximum temperature by the materials used in its construction, it is seen that this effectively determines the maximum rating of the machine.

Since the machine losses play a key role in determining a machine rating, the nature of these losses merits further consideration. Figure 2-11 shows the types of power loss that are commonly present in a dc machine. These losses will vary greatly with respect to each other and may be the predominant loss component, or may be negligible, depending upon the detail of design. An important facet of machine design is to be able to accurately predict the various losses for a particular design and to then manipulate design parameters in such a way as to minimize the total.

The loss components and factors which influence them are as follows:

Copper loss. This is usually a principal loss component in a dc machine and it includes all types of joule heat generation. This type of loss is generated by current flow and the resistance of the copper to it. It is characterized as an I^2R loss, i.e., it is proportional to a resistance and the square of a current. It can exist in a number of forms which are not always obvious but may nevertheless be significant. Among the forms of I^2R loss which are likely to be present are:

a. Load current loss is the most obvious form of I^2R dissipation. This loss component is determined directly by the product of the armature resistance and the square of the armature current. Since armature current increases with

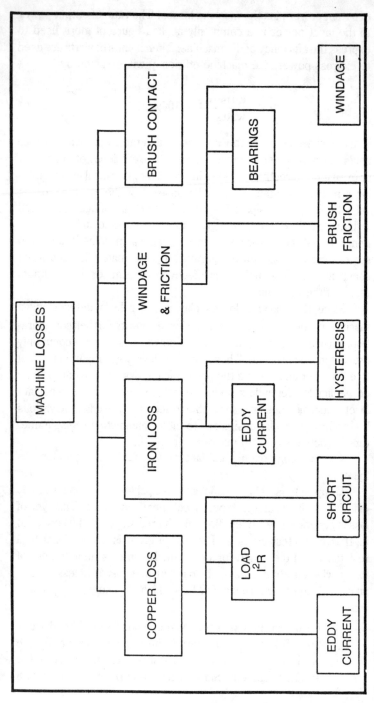

Fig. 2-11. Breakdown of power losses in a dc machine.

load torque, it can be seen that this form of heating increases rapidly with load. For a given current requirement, it would then be advantageous to wind the armature slots with as large a wire as possible in order to minimize the armature resistance.

b. Short-circuit losses are another form of I^2R dissipation which occurs in the armature coil that is undergoing commutation. The existance of this current can be seen by referring to Fig. 2-9(b). At the moment that the brush contacts more than one commutator segment it provides a path for a current to circulate through the coil. This condition is shown in the developed winding diagram of Fig. 2-12. The short-circuit current is influenced heavily by the contact drop of the brush which is usually large compared to the impedance provided by the coil itself. Another factor influencing short-circuit currents is the ability of the motor to commutate, or quickly reduce stored energy to zero. Thus, factors which promote the existence of induced voltages in the commutating coil will also increase this loss component.

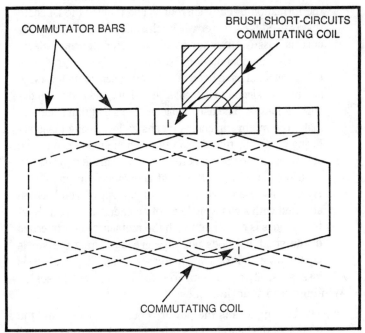

Fig. 2-12. The short-circuit current in a commutating coil.

c. Eddy currents are usually not a significant form of I^2R dissipation in armatures that are wound with magnet wire for medium to high voltages. However, they can become significant in motors whose armatures are built with conductors of large cross section, often of rectangular shape. This type of construction is common to traction motors built to operate off relatively low battery voltages. In cases where this form of loss becomes excessive it might be necessary to laminate the copper strap used for armature conductor in order to increase the resistance to eddy currents. Eddy-current losses in the armature conductors are influenced by the air-gap magnetic field strength and by the resistance of the conductor to the formation of eddy currents.

Iron Loss. The total iron loss consists of two components:

a. Eddy-current loss in the steel has the same characteristics as that previously mentioned as a form of copper loss. It is for the purpose of minimizing eddy-current losses that machines are built from laminated iron cores. The use of laminates which are insulated from one another minimizes the formation of eddy currents. The thinner the laminate that is used the greater will be the effect on reducing eddy current losses. Another steel characteristic that affects this loss component is its resistivity. Special grades of electrical steel have been developed which contain varying amounts of silicon to increase resistivity and reduce eddy-current losses. The use of silicon steels and thin sheets to make the motor punchings means an increasing manufacturing cost. For that reason the design engineer must be able to estimate the expected losses and make good design decisions based upon the most economical design.

b. Hysteresis loss is the second component of iron loss associated with a changing level of magnetic induction. Hysteresis loss is associated with the metallurgical properties of the steel and does not vary greatly among the steels commonly used for core materials. Similarly, hysteresis loss is usually small compared to eddy-current losses.

Windage and Friction Loss.

a. The bearing loss is the power required to overcome the friction of the bearings. For a ball bearing, this loss component would be very small unless some type of sealed bearing were used which usually significantly increases the bearing loss. Bearing loss tends to remain approximately constant over a speed range, i.e., coefficient of friction decreases with an increasing speed.

b. The windage loss is the power required to rotate the motor minus the bearing loss. Windage loss is heavily influenced by whether or not the motor has an internal ventilating fan. If the motor is without a fan, the windage loss will be small compared to the copper and iron losses. If the motor does have an internal fan, though, the windage loss can be very significant and will increase approximately in proportion to the cube of the speed.

c. Brush friction loss is due to the drag of the brushes against the commutator. This loss component is largely determined by the magnitude of the spring pressure used with the brushes. It will also be influenced by the coefficient of friction between the particular brush material and commutator used in a specific design. Brush friction is at a maximum when the motor is at a standstill. As the motor comes up to speed, the friction between brush and commutator decreases. The decrease is due in part to brush bounce, aerodynamic lift of the brush, and the effect of sliding friction. Brush friction loss is usually a small part of the total machine loss.

Brush Contact Loss.

The brush contact loss is the loss associated with the passing of current through the brush and commutator interface. The loss is equal to the product of armature current and the contact drop in volts. This component of motor loss is often significant in reducing the overall efficiency of the motor. The selection of brushes with a high contact drop is often resorted to when a motor is subjected to difficult commutation conditions. The high contact drop characteristic aids in commutation but the price that is paid is higher loss.

The preceding paragraphs are a brief discussion of the main loss components present in an electric machine. The total effect of these losses determines the machine efficiency, its resulting temperature rise, and consequently the machine rating. More detailed discussion of several of the machine losses will be made in a subsequent section where specific machine conditions are dealt with.

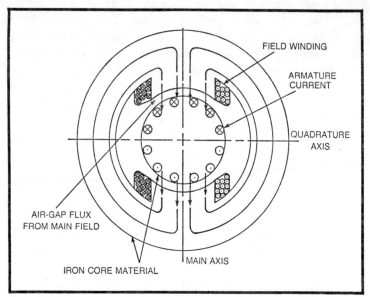

Fig. 2-13. The functional parts of a dc machine.

MAGNETIC FIELDS IN THE DC MACHINE

All dc machines are characterized by the presence of two magnetic fields during operation. The two fields are:

1. the main field established by the field coils and the field current.
2. the armature reaction field which results from the armature current flowing through armature conductors.

The field coils are wound around a salient pole, which is a part of the stationary motor structure (Fig. 2-13). The function of the field winding is to create magnetic lines of force (flux) in the air gap between the armature and the field pole face. This air-gap flux becomes the medium for the development of torque and subsequent energy conversion. Since the equations for calculation of voltage and torque both contain the magnetic flux per pole, it is readily apparent that a principal design consideration is to provide as much flux per pole as possible. For that reason, machines are designed with relatively small clearances between the rotor and stator structures in the sector around the main axis.

The armature reaction field axis is determined by the position of the brushes (Fig. 2-13), and occurs at that orientation where armature coil currents are switched or commutated. If the brushes are adjusted to provide equal characteristics for both directions of rotation, the armature reaction will be at a right angle to the main field or on the quadrature axis. Sometimes the armature reaction is represented as a single vector acting along the quadrature axis as shown in Fig. 2-14. While this representation is adequate for the purpose of illustrating a concept it should be kept in mind that the armature reaction has a distribution around the entire periphery of the rotor that is determined by the individual coil locations.

The effects of armature reaction are generally undesirable. In particular, a high level of armature reaction contributes to commutation difficulties. For this reason it is common design practice to have very large air gaps on the quadrature axis (Fig. 2-13) so as to minimize magnetic flux in this interpole area where commutation occurs. Another effect of armature reaction is to demagnetize the air gap. This demagnetizing effect is illustrated in Fig. 2-14. It is seen that components of armature reaction do exist which oppose the main field flux over half the pole face sector, finally reducing to zero on the main axis. At light loads, when the armature current is relatively small, the demagnetization will be negligible. As load torque on the motor is increased, however, armature current also increases, so that motors having a fixed field strength will experience significant demagnetization and resulting loss of torque developing capability.

The function of the main magnetic field may be provided by electromagnets as in the figures shown. Or it may also be provided by replacing the salient pole and field assembly (electromagnet) with a permanent magnet as shown in Fig. 2-15. Whether the magnet

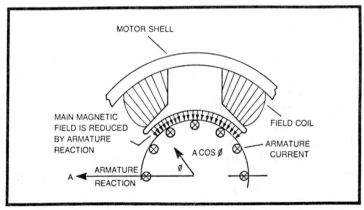

Fig. 2-14. Armature reaction and its demagnetizing effects.

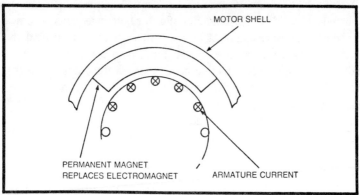

Fig. 2-15. DC machine with a permanent-magnet field.

field is provided by a coil winding or a permanent magnet, all of the effects which have been discussed are equally valid.

SUMMARIZATION

The dc machine of today is based upon a well developed theory that had its origin in 1820 with the discovery of electromagnetic forces. It is possible to accurately predict machine performance using formulae derived directly from equations initially established by Ampere. These formulae can be applied to either generator design or motor design with the same degree of confidence. The principle of energy conversion is the same for generator or motor action, the only difference is the direction of energy flow.

Conventional dc machines utilize a mechanical device called a commutator to change direction of current flow in the armature conductors. This switching or commutation is necessary in order that the motor (or generator) provide a steady unidirectional output.

Energy is conserved in an electric machine. The total electric power is equal to the machine output plus all the power losses in the machine for motor action. While in the generating mode of operation, the mechanical input will be equal to the electrical output plus all of the various losses. The machine losses occur in a variety of ways including electrical, magnetic, and mechanical effects.

The dc machine has two identifiable magnetic fields. One of the fields, the main field, constitutes the medium by which energy conversion occurs. Machine design is oriented to maximize the main field. The second field is associated with armature current and often produces undesirable effects. Machine design attempts to minimize the strength of this field.

Chapter 3
Constructional
Features of DC Machines

All dc machines are made from component parts which must provide tightly prescribed functions. The component parts may vary greatly in appearance but the function is unvarying. Despite the great disparity in physical appearance, the function of a motor part is usually readily identifiable upon inspection. The main functional parts of a dc machine will always include the following:

1. There must be a part of the machine that functions to establish a strong magnetic field. From previous sections of the book it is recognized that the field is the necessary medium for energy conversion.

2. There must be an arrangement of conductors interconnected in such a way so as to allow current flow. The conductors must be able to move relative to the magnetic field. This assemblage of conductors is called the armature.

3. There must be a form of commutator which acts to reverse the current flow through the armature conductors as they move from a field of one polarity to that of the opposite polarity. The commutation may be accomplished by mechanical means, as with conventional brushes and commutator, or it may be done using solid state switching devices.

4. Finally, there must be a mechanical structure to hold the necessary functions together in a proper manner. The

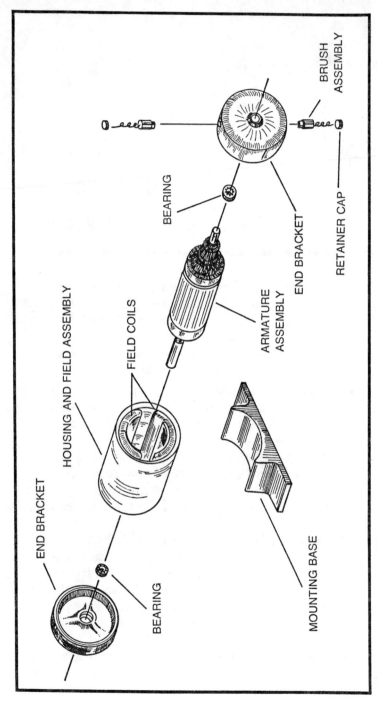

Fig. 3-1. Exploded view of a dc machine.

structure must provide a bearing and shaft arrangement that allows the relative movement between the armature and the field. It must be adequate to withstand mechanical stresses, provide heat dissipation, protect from hostile environments, and protect against electrical hazard.

When all of these functional requirements have been provided for, the result will be a dc machine capable of converting energy. A typical machine assembly embodying all of the above functional parts is shown in an exploded view in Fig. 3-1. It can now be appreciated that the essence of motor design involves the achievement of the necessary functions at a minimum cost of labor and material so as to produce the desired machine output. This chapter will discuss in detail the necessary parts of a dc machine. Included in the treatment will be a description of how and why parts are made as they are. Figure 3-2 shows a cutaway view of a typical dc motor.

THE MAGNETIC CIRCUIT IN A DC MACHINE

In order to establish a strong magnetic field in a machine, it is necessary to use ferrous materials (iron, steel) for the cores around which the electrical coils are placed. The function of the steel core is to provide a path that is easily penetrated by the magnetic lines of

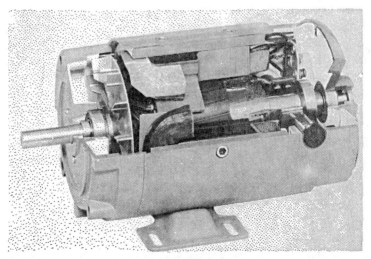

Fig. 3-2. A view of a dc motor with a portion machined away to disclose the interior parts. (Courtesy Gould Inc., Electric Motor Division.)

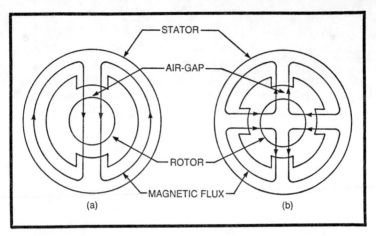

Fig. 3-3. Magnetic circuits in dc machines, (a) two poles and (b) four poles.

force. This property of a material is called its permeability. Often-times permeability of a material is given as a relative permeability by comparing it to that of air, which has a permeability of one. Using relative permeability as an index of their ability to carry magnet flux, it is found that most steels have permeabilities ranging up to the thousands. An electromagnet made with an iron core thus has several thousand times more magnetic flux than it would if some other core material were used such as aluminum.

It is necessary to design a dc machine so that there is a complete magnetic circuit through the structure. In order to achieve a strong magnetic field, and a high level of output, it is necessary to minimize the amount of air gap through which the magnetic flux must pass. For this reason machines are usually designed for a minimum air gap that is consistent with other mechanical limitations. For example, it would be a poor design to make an air gap of such a size where mechanical vibrations could result in striking between rotating and stationary parts. For this reason air gaps will range from a minimum of about 0.015 inch in small precision made motors to more than 0.25 inch in very large machines.

Typical machine magnetic-circuit design is shown in Fig. 3-3. While two-pole configurations are shown, any even number of poles can be used. The figure also illustrates a feature that is common to all electric machines, ac as well as dc. There is a stationary portion called a stator which usually, but not always, supports the field winding. There is a second portion that is free to rotate in a set of bearings; this part is called the rotor. The rotor usually has the armature winding positioned upon it.

50

THE FIELD ASSEMBLY

It is worth noting that the terms stator and field assembly are often used interchangeably, and with some confusion when applied to dc machines. The stator is the structural part of the motor that remains stationary during operation. On the other hand, the field assembly is the part of the motor that creates the magnetic field. Now while the field assembly is always stationary in the case of conventional dc machines, it does not necessarily have to be. It will be seen in a later section that in some types of brushless motors the field assembly rotates as part of the rotor while the armature is placed upon the stator and remains fixed.

This section, however, is concerned with a description of conventional motor design. With this qualification the field assembly can be seen to consist of the following parts (Figs. 3-4 and 3-5):

1. A core or pole piece made of ferrous material to establish the magnetic field.
2. A coil winding to provide the exciting current for the main field.
3. Insulating materials between the coil and the core to preserve the winding and protect the motor user against electrical shock.

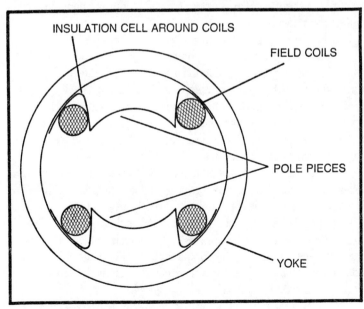

Fig. 3-4. The field winding assembly of a dc machine.

Fig. 3-5. Pole piece and coil assembly. (Courtesy Gould Inc., Electric Motor Division.)

    4. A yoke structure to hold the pole pieces and to complete the magnetic circuit from one pole to another.

In many machines it is not necessary from the standpoint of controlling iron losses to fabricate the pole piece from laminated steel sheet. This is because the magnet flux in the pole piece stays at an essentially constant value. This being the case the pole piece may be made from sheet steel as a matter of production convenience and manufacturing cost.

The field coils are wound from film insulated wire or in some cases copper strap depending upon the current capacity that is required. Commonly used production practice is to first wind the coil around a form which has the same size and shape as the pole piece. After the coil is so formed it will be dipped in varnish and baked to fill the wire interstices and prevent moisture formation in the coils. After the coils are formed they are placed around the pole piece for subsequent assembly operations.

An alternate method of winding field coils is commonly used on smaller machines. This technique is to utilize specially designed automatic machines that wind the coils directly around the insulated pole pieces after they have been put into place in the yoke. This type of production technique is limited to motor designs where the production volume will justify the capital investment required by high productivity equipment. Automatic winding equipment also has a limitation in not being able to handle large wire.

The insulating material used between the field coil and pole piece is of critical importance in machine design. It is not uncommon in the case of wound field motors for the motor hot spot to develop in the area between the field coils and the core. Since the maximum allowable temperature is limited according to the class of insulation material used, the insulation in effect determines the rating of the machine. Class B insulation systems are in most common usage for industrial quality of motors, with class F systems also frequently encountered. Class H systems are much less frequently used on dc machines than they are on the ac machine counterparts.

The final part of the field assembly is the yoke. In many small integral hp dc machines (under 30 hp) the function of the yoke and motor housing are incorporated into one part. In this type of design the yoke is usually made from a piece of steel tubing. The thickness of the tubing is determined by its capacity to carry the necessary magnetic flux. This type of design makes for a very robust mechanical structure since the tube thickness required for magnetic flux is usually several times that required for mechanical strength. Such a field assembly design is often used for traction motors, which are subjected to extreme mechanical stresses (Fig. 3-6).

THE ARMATURE ASSEMBLY

In conventional dc motor construction the armature winding is always associated with the rotating part (rotor) of the motor. As a result the term "armature," in its common usage, is the more frequently used name for the rotating member of the motor (Fig. 3-7).

Fig. 3-6. Stator and rotor assemblies from a traction motor. (Courtesy Gould Inc., Electric Motor Division.)

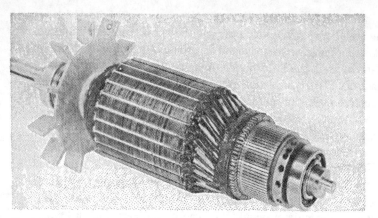

Fig. 3-7. Armature/rotor assembly of a dc motor. (Courtesy Gould Inc., Electric Motor Division.)

The function of the armature in the energy conversion process is to carry the electric current (load current) which in turn reacts with the magnetic flux of the main field. As was shown in Chapter 2, both voltage generation and torque development are associated with energy conversion. The armature winding and its connection to the commutator are the key factors in determining a machine's operating characteristics. Thus the armature assembly (rotor) size and the maximum allowable temperature will determine the voltage and current characteristics. In other words, a one horsepower motor will be the same size regardless of whether it is rated for 200 volts or 48 volts. The 200 volt armature, though, will have many turns of small wire size while the 48 volt armature will have fewer turns of a larger wire size.

Although the functions of armature conductors and the commutator are two distinctly separate things, in practice the commutator is regarded as an integral part of the armature assembly. This is because after connecting the armature windings to the commutator, the commutator has, in fact, become a part of the armature circuit. With this thought in mind, the parts of the armature assembly can be described as follows:

1. There is an armature core made of steel to support a strong magnetic field. It is very important that the armature core be made of steel laminations in order to minimize hysteresis and eddy-current losses. These steel laminations can be made of varying thicknesses and from a variety of steel grades. The choice of steel grade and the sheet

thickness become an economic trade off between considerations of manufacturing costs and required motor performance. Core losses do decrease when thinner laminations are used, but the manufacturing cost will increase, as a rule. As can be seen in Fig. 3-8, the core laminations are punched with deep slots to hold the armature conductors.

2. The second essential part of the armature assembly is the winding itself. As was mentioned previously, the number of turns and wire size determines the voltage and current characteristics of the motor. The armature may take on a variety of appearance depending upon the intended use of the machine. Figures 3-9 and 3-10 show respectively the armatures from a traction motor, and from a low inertia servo motor. The traction motor armature is excited with a relatively low value of voltage from an industrial battery. Consequently, the armature utilizes only one conductor in each coil side. But the motor is required to develop a substantial amount of power. The result is that a large conductor size is required to handle the necessary input current.

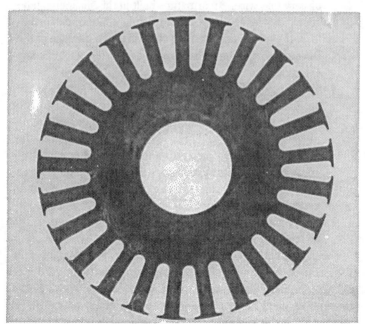

Fig. 3-8. The rotor lamination as punched from sheet steel. (Courtesy Gould Inc., Electric Motor Division.)

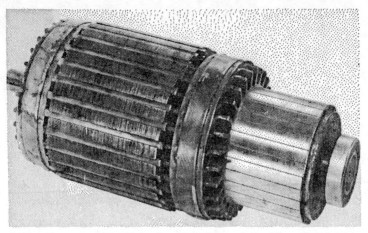

Fig. 3-9. Rotor assembly of a traction motor. (Courtesy Gould Inc., Electric Motor Division.)

In contrast, the small servomotor armature shown in Fig. 3-10 is not required to handle a significant power; its purpose is to provide a high acceleration. For such an application the armature coils are wound with many turns of very fine wire. An armature built in this manner can be seen to be very economical of material content. Nevertheless, the total manufacturing costs are high because of the large amount of hand labor required and a high rate of scrap loss in production.

3. The commutator becomes an integral part of the armature assembly after it is connected to the conductors. The commutator segments are usually made of copper with a minute percentage of silver also present to stabilize the physical properties at high temperatures. Commutators come in a wide variety of shapes and sizes. The armatures shown in Figs. 3-9 and 3-10 exemplify the most commonly used commutator type. The disc commutator is less commonly used, seeing greatest application with very small instrument type motors.

The process of building an electric machine can be regarded as starting with the shaft. The shaft is invariably turned from hardenable alloys especially where substantial torque loads will exist. Many of the smaller sized motors, especially those intended for military application or servo systems where corrosion is undesirable, are built with stainless steel shafts. However, this feature becomes very

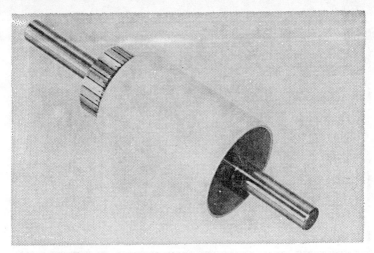

Fig. 3-10. Rotor from a low inertia servomotor.

costly with increasing size and is rarely seen in general purpose motors above one horsepower.

After the armature core punchings and the shaft are available, the next step is to push the laminated core onto the shaft. Such an assembly is shown in Fig. 3-11. The laminations are secured tightly to the shaft by making an interference fit between the bore dimension of the laminated core and the shaft diameter. The security of this fit is often additionally reinforced by raising a portion of the shaft surface by knurling it or staking it at several points.

After the core is secured to the shaft it must be insulated to protect the windings which are to be placed upon it. The most common technique used to insulate the armature core slots utilizes sheet insulation. The sheet insulation is cut to size and then shaped

Fig. 3-11. Armature (rotor) core pressed on the shaft. (Courtesy Gould Inc., Electric Motor Division.)

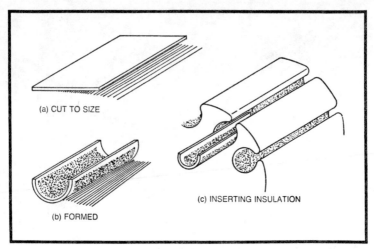

(a) CUT TO SIZE

(b) FORMED

(c) INSERTING INSULATION

Fig. 3-12. The use of a slot cell to insulate the electrical winding from the core.

to conform to the slot in which it is to be inserted (Fig. 3-12). This type of insulation is called a slot cell. It protects the portion of the coil lying in the core slot from electrically grounding to the core. Another piece of insulation is punched using a die similar to the one used to punch the steel laminations. This insulation is used at either end of the armature core to protect the coils where they emerge from the slots and bend around the corner of the core. (Fig. 3-13). After the core insulation is in place, the armature coils are put into place and the slot opening is usually closed with a wedge, also made from an insulating material. The electric coil sides are then completely encased and protected from the armature core, an important safety feature in protecting against electrical shock.

An alternate method of insulating armature cores is commonly used in machines of fractional and subfractional horsepower sizes. This technique employs an epoxy applied directly to the surface of the core. This process involves preheating the cores to a temperature above the fusing temperature of the epoxy that is to be used. A finely granulated epoxy powder is then introduced to the core by means of an air stream. The air stream distributes the epoxy particles uniformly over the core surface where they adhere and fuse, covering the entire surface. After the coated epoxy core has been given a heat treatment to cure the epoxy, it is provided with a very tough insulation on all exposed surfaces. The epoxy insulated core offers an advantage of good heat transfer, good mechanical strength, and low manufacturing costs. The main disadvantage is the difficulty in obtaining a uniform quality of insulation on all surfaces where it is required. Special production equipment has been developed that uses an electrostatic field around the core when the epoxy spray is applied. The electrostatic field attracts the particles to the critical areas of the core and helps to produce consistently good quality.

As is the case with the field assembly, the material used to insulate the armature is also critical in determining a machine's rating by establishing a maximum allowable temperature. The design of the armature insulation system is usually a difficult balance between the demands for electrical insulation and the need for heat transfer. For as power is dissipated in the armature coils, it must be gotten rid of in order to avoid an excessive temperature rise. In the armature assembly a good part of the joule heat produced in the coils is transferred by conduction through the slot cell insulation to the core. Unfortunately, materials that have good electric insulating properties have very poor thermal conducting properties. For this reason it is usually desirable to control the thickness of the insulation; too thick an insulator contributes to a hot motor while too little insulation can cause a safety hazard or machine failure.

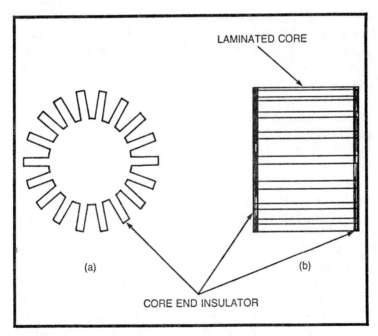

Fig. 3-13. End insulation of the armature core, (a) the punched insulation and (b) the insulation in place at either end of the core.

The armature coils may be wound around a form to obtain the right shape and then placed by hand into the core slot, or they can be wound directly upon the insulated core. Usually larger sized machines are built using hand methods while smaller machines use an automatic or semiautomatic method. If the production volume justifies the investment, it is possible to obtain armature winding equipment that is fully automatic. This equipment can put the winding on the core at rates of hundreds of turns per minute and automatically index from one slot to the next when the required number of conductors are in place. Whenever automatic equipment can be used, it is very effective in increasing productivity and reducing manufacturing unit costs.

After the coils are in place, the coil ends must be connected to the commutator. The connection may be made in a number of different ways. It may be done by soldering, brazing, welding, or fusing. The important thing is to obtain a good electrical connection. The method of connection is often dictated by the characteristics of the commutator and conductors and the equipment that is available. Production equipment is available for automatically making all the required connections using the fusing technique. In this technique one electrode of the machine is placed upon the commutator bar. The second electrode, of opposite polarity, is forced down upon the conductors to be connected to that bar. While maintaining the pressure, an electric current of thousands of amps at several volts potential passes between the electrodes. As a consequence of the great heat created at the point of contact between the conductor and commutator bar, a fusing of materials is achieved. The result is a very good and reliable electrical connection.

Upon making the armature-commutator connections it is common practice to apply an electrical grade of varnish to the entire assembly. The varnish flows through the coils, filling the winding interstices and fusing individual conductors into a solid mass. The varnish functions to prevent movement of the wires in vibration modes, which could lead to a failure over a long period of time. It also acts to exclude moisture formation in the coils which could lead to a dielectric breakdown.

The next operation on the armature assembly is the truing of the commutator surface. In order that the brushes maintain good electrical contact with the commutator surface during running conditions it is necessary that the commutator diameter be concentric with the shaft axis of rotation. To accomplish this the armature assembly is mounted in a lathe between centers and the commutator turned. Special attention must be given to the commutator surface finish. If the final cut on the commutator is too heavy and the tool feed too fast, a rough surface may be left which will cause brush contact problems during the operation of the machines. To provide a good surface finish and minimize the potential for brush contact problems a diamond cutting tool is often used. The diamond tool cuts the commutator copper with much less burr than a conventional steel cutting tool. The result is a mirror like finish on the commutator. Good commutator finishes are often specified in the range of 30 to 100 microns. After the turning operation, it is necessary to clean the slots between the commutator segments. This is done to make sure that machining burrs do not bridge the insulation slot and short circuit an armature coil. The slot cleaning operation will also make sure that commutator insulation is below the brush diameter where it cannot interfere with the electrical contact.

The final operation on the armature assembly is a balancing operation. It is almost impossible to avoid some slight dissymmetry from occurring in the process of manufacturing a complete rotor. Therefore, in order to avoid undesirable vibrations which could occur at high rotational speeds, the finished armature assembly is balanced in a dynamic mode. The balancing is done by turning the rotor at a specified speed while supporting it upon the bearing journals of the shaft. The balance machine elements supporting the shaft contain piezoelectric crystals which convert mechanical vibrations to proportional electric voltages. The machine thus provides a means of qualifying the unbalance in a rotor. A strobe light triggered by the unbalance also provides an indication of the angular sector where the unbalance exists. Compensation is then provided for the unbalance by either adding or taking away mass from the appropriate sector of the armature assembly. The balancing operation is somewhat of a trial and error process which continues until the amount of vibration falls within the specified limits. Production rotors are usually tested by comparing their degree of unbalance to that of a standard rotor that gives acceptable vibration in the finished motor.

The armature assembly in Fig. 3-7 shows a balance disc between the commutator and bearing. The holes in the disc are evidence of the mass removal to balance the assembly.

THE MECHANICAL STRUCTURE

The mechanical structure of an electric machine takes on an infinite variety of sizes and shapes. There has been an attempt to

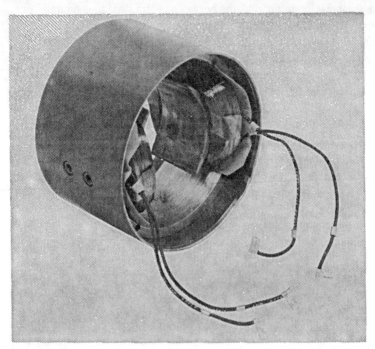

Fig. 3-14. Shell and stator construction showing thin wall shell design concept. (Courtesy Gould Inc., Electric Motor Division.)

standardize those features that affect the interchangeability of machines. The result is the National Electric Manufacturers Association (NEMA) standards which specify frame sizes, shafting, and mounting information. But even motors built to the NEMA standards show wide variations from one manufacturer to another. In addition, there is a very substantial portion of the dc machinery market where no attempt is made to adhere to a standard motor configuration. This is especially true of the markets for traction motors and servo motors where each customer wants a motor especially designed for his particular application. This insistence upon specialty motor features is a principal cause for high manufacturing costs and resulting high price for the motor user.

The main elements of the mechanical structure are the shell, or housing (Fig. 3-14), and the end brackets. A number of shell design options are available depending upon the design of the field assembly. These design options include:

1. The shell may be made of a relatively thin walled material if a separate magnetic yoke is included in the field assembly.

If this design concept is followed, the shell may be made from a nonferrous metal such as aluminum. In this case the shell is strictly a structural part that holds the machine together. The separate yoke serves the function of carrying the magnetic lines of force necessary for energy conversion.

2. The shell wall may be utilized as an integral part of the magnetic circuit. In such a case the shell wall must be thick enough to carry the required air-gap flux (Fig. 3-2) and must be fabricated from a magnetic material.

3. The shell may be made from laminated sheet steel (Fig. 3-15). This type of construction is fairly common with small fractional hp motors where economy of manufacture is the primary design criterion. In such design the function of the field assembly core and structural support are combined into a single part.

Fig. 3-15. Unitized stator/field lamination design. (Courtesy Gould Inc., Electric Motor Division.)

The function of the end brackets is to provide a stable support for the rotor bearings and to hold the machine together. In addition, the end bracket at one end of the machine must provide a means of supporting the brushes, which must provide electrical contact to the commutator.

Different features incorporated into the end-bracket design are among the most frequently encountered machine variations. Brackets can be provided with ventilating holes for machines where the application allows the use of an open frame. Or, totally enclosed brackets are also available if such is required by the installation.

The matter of an open or enclosed frame is a very important consideration in the selection of a proper machine. The use of an open frame with an internal ventilation fan can quadruple the heat dissipation of a machine. The improved heat dissipation has the effect of significantly increasing the horsepower rating of a given size of machine.

The end brackets can be made in a variety of ways. In small motor sizes, where production volumes are large, aluminum die-cast parts will likely be seen. An aluminum die-cast part provides close mechanical tolerances on dimensions and superior heat transfer characteristics at a low manufacturing cost. Other frequent bracket variations may include different mounting face configurations, tapped holes for mounting accessories, etc. Where production volume is small, sand-cast or investment-cast parts can be used without making the capital investment in an expensive die-cast mold. Cast iron end brackets are most common on larger machines and where an extra degree of mechanical strength is needed, as for traction motor application.

THE BRUSH SUPPORT MECHANISM

The design of the brush support mechanism is often critical to the successful operation of a dc machine (Fig. 3-16). In order for good machine operation to occur it is necessary that the brushes maintain an intimate contact with the commutator surface at all times. Achieving this intimacy is not always as easy as it may seem at first glance. For it must be remembered that the commutator is often turning at a high rotational speed. If there is any error in the concentricity of the commutator surface it tends to impart a radial acceleration to the brush which causes it to move away from the commutator surface. Of course if brush movement is sufficient, electric current flow will be interrupted. The result is an electric

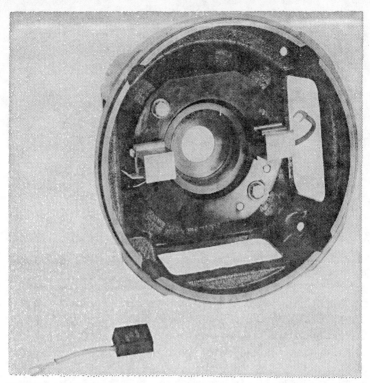

Fig. 3-16. Brush holder arrangement inside the motor and bracket. (Courtesy Gould Inc., Electric Motor Divison.)

sparking under the brushes which can drastically increase the rate at which the brush wears, shortening the useful life of the machine.

Because of the very deleterious effects of sparking, brush support design is given a great deal of attention. The brush is invariably spring loaded with the spring pushing the brush against the commutator with a positive force. This force must be designed to withstand the accelerating forces which act to make the brush bounce and interrupt the current flow. The spring also provides the function of feeding the brush against the commutator as the brush wears away. In the ideal case the spring should maintain a constant force against the brush as it wears away to keep it at an optimum pressure against the commutator. To achieve this result special constant force springs are often used. The brush pressure is also a factor in determining the contact drop at the brush.

A number of brush-holder designs are shown in Fig. 3-17. It is very important that this portion of the brush support mechanism be

Fig. 3-17. Various types of brush-holder support mechanisms. (Courtesy Gould Inc., Electric Motor Division.)

rigid and sturdy enough to withstand mechanical stresses. There are many applications subjecting motors to abusive treatment, which may include vibrations and shock. If a brush-support mechanism deflects under such rough treatment, however slight the deflection might be, it will likely contribute to problems of short brush life. A well designed brush-holder mechanism will therefore accurately locate the brushes in the neutral or commutating zone while at the same time providing a very solid and stable support. It is common practice for the dc machine to incorporate a constructional feature that allows the brush holders to be rotated through a limited angle. This feature allows adjustment to be made to the brush-holder position by turning it to the neutral axis.

THE CLASSIFICATION OF MOTOR ENCLOSURES

Standards have been developed to classify the type of motor enclosure according to the particular cooling and/or environmental conditions of its application. The enclosure classes are as follows:

Drip-Proof Machine

A drip-proof machine is an open machine in which the ventilating openings are so constructed that successful operation is not interfered with when drops of liquid or solid particles strike or enter the enclosure at any angle from 0 to 15 degrees downward from the vertical.

Splash-Proof Machine

A splash-proof machine is an open machine in which the ventilating openings are so constructed that successful operation is not interfered with when drops of liquid or solid particles strike or enter the enclosure at any angle not greater than 100 degrees downward from the vertical.

Semiguarded Machine

A semiguarded machine is an open machine in which part of the ventilating openings in the machine, usually in the top half, are guarded as in the case of a "guarded machine" but the others are left open.

Guarded Machine

A guarded machine is a open machine in which all openings giving direct access to live metal or rotation parts (except smooth rotating surfaces) are limited in size by the structural parts or by screens, baffles, grilles, expanded metal, or other means to prevent accidental contact with hazardous parts. Openings giving direct access to such live or rotating parts shall not permit the passage of a cylindrical rod 0.75 inch in diameter.

Drip-Proof Fully Guarded Machine

A drip-proof fully guarded machine is a drip-proof machine whose ventilating openings are guarded in accordance with the requirement of a guarded machine.

Open Pipe-Ventilated Machine

An open pipe-ventilated machine is an open machine except that openings for the admission of the ventilating air are so arranged that inlet ducts or pipes can be connected to them. This air may be circulated by means integral with the machine or by means external to and not a part of the machine. In the latter case, this machine is sometimes known as separately or forced-ventilated machine. Enclosures may be as defined in the previous five machines.

Open Externally Ventilated Machine

An open externally ventilated machine is one that is ventilated by means of a separate motor-driven blower mounted on the machine enclosure. Mechanical protection may be defined as in the first five machines. This machine is sometimes known as a forced-ventilated machine.

Weather-Protected Machine

Type I. A weather-protected Type I machine is an open machine with its ventilating passages so constructed as to minimize the entrance of rain, snow, and airborne particles to the electric parts and having its ventilated openings so constructed as to prevent the passage of a cylindrical rod 0.75 inch in diameter.

Type II. A weather-protected Type II machine shall have, in addition to the enclosure defined for a weather-protected Type I machine, its ventilating passages at both intake and discharge so arranged that high velocity air and airborne particles blown into the machine by storms or high winds can be discharged without entering the internal ventilating passages leading directly to the electric parts of the machine itself. The normal path of the ventilating air which enters the electric parts of the machine shall be so arranged by baffling or separate housings as to provide at least three abrupt changes in direction, none of which shall be less than 90 degrees. In addition, an area of low velocity not exceeding 600 feet per minute shall be provided in the intake air path to minimize the possibility of moisture or dirt being carried into the electric parts of the machine.

Totally Enclosed Nonventilated Machine

A totally enclosed nonventilated machine is a totally enclosed machine which is not equipped for cooling by means external to the enclosing parts.

Totally Enclosed Fan-Cooled Machine

A totally enclosed fan-cooled machine is a totally-enclosed machine equipped for exterior cooling by means of a fan or fans integral with the machine but external to the enclosing parts.

Explosion-Proof Machine

An explosion-proof machine is a totally-enclosed machine whose enclosure is designed and constructed to withstand an explosion of a specified gas or vapor which may occur within it and to prevent the ignition of the specified gas or vapor surrounding the machine by sparks, flashes, or explosions of the specified gas or vapor which may occur within the machine casing.

Dust-Ignition-Proof Machine

A dust-ignition-proof machine is a totally enclosed machine whose enclosure is designed and constructed in a manner which will exclude ignitable amounts of dust or amounts which might affect performance or rating, and which will not permit arcs, sparks, or heat otherwise generated or liberated inside of the enclosure to cause ignition of exterior accumulations or atmospheric suspensions of a specific dust on or in the vicinity of the enclosure.

Waterproof Machine

A waterproof machine is a totally enclosed machine so constructed that it will exclude water applied in the form of a stream from a hose, except that leakage may occur around the shaft provided it is prevented from entering the coil reservoir and provision is made for automatically draining the machine. The means for automatic draining may be a check valve or a tapped hole at the lowest part of the frame which will serve for application of a drain pipe.

Totally Enclosed Pipe-Ventilated Machine

A totally enclosed pipe-ventilated machine is a totally enclosed machine except for openings so arranged that inlet and outlet ducts or pipes may be connected to them for the admission and discharge of the ventilating air. This air may be circulated by means external to and not a part of the machine. In the latter case, these machines shall be known as separately ventilated, or forced-ventilated machines.

Totally Enclosed Water-Cooled Machine

A totally enclosed water-cooled machine is a totally enclosed machine that is cooled by circulating water, the water or water conductors coming in direct contact with the machine parts.

Totally Enclosed Water-Air-Cooled Machine

A totally enclosed water-air-cooled machine is a totally enclosed machine that is cooled by circulating air which, in turn, is cooled by circulating water. It is provided with a water-cooled heat exchanger for cooling the ventilating air and a fan or fans, integral with the rotor shaft or separate, for circulating the ventilating air.

Totally Enclosed Air-to-Air-Cooled Machine

A totally enclosed air-to-air-cooled machine is a totally-enclosed machine which is cooled by circulating the internal air through a heat exchanger which, in turn, is cooled by circulating external air. It is provided with an air-to-air heat exchanger for cooling the ventilating air and a fan or fans, integral with the rotor shaft or separate, for circulating the internal and a separate fan for circulating the external air.

Totally Enclosed Fan-Cooled Guarded Machine

A totally enclosed fan-cooled guarded machine is a totally enclosed fan-cooled machine in which all openings giving direct access to the fan are limited in size by the design of the structural parts or by screens, grilles, expanded metal, etc., to prevent accidental contact with the fan. Such openings shall not permit the passage of a cylindrical rod 0.75 inch in diameter.

Chapter 4
Types of DC Machines

From the earlier discussion of Chapter 2 it can be seen that there are two essentials for electromechanical energy conversion. One of these essentials is the existence of a magnetic field which becomes the medium for force development. The second essential is the controlled flow of electric current in the magnetic field medium. As long as these two essentials exist, motor or generator action can proceed. As has also been mentioned earlier, machine design focuses upon the optimization of these two factors with the least expenditure of material and labor cost.

While all electric machines will have similar functional parts, they are also available in a large variety of forms. These different types of machines may be obtained by the use of different connections of the machine windings. As will be seen in this section, the manner of connection of field and armature windings can greatly influence a machine's output characteristics as a function of its speed.

A second type of machine variation is obtained by mechanical and electrical design which emphasizes the optimization of machine parameters for special application. Examples of this type of variation can be seen in the design of servomotors where the minimization of rotor inertia is a paramount application consideration. The attainment of special machine characteristics often results in machines of extraordinary appearance. It should always be kept in mind, however, that the parts of a machine are common in function for all types regardless of their application.

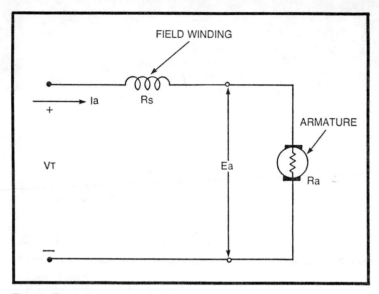

Fig. 4-1. The series connected motor.

In this chapter the most commonly encountered types of machines are discussed.

THE SERIES-CONNECTED-FIELD MOTOR

The series motor is one in which the motor field winding is connected in series with the armature. The combination of windings is then connected across the supply voltage. The schematic for the series motor is shown in Fig. 4-1. With this connection it can be seen that the current which flows through the armature conductors is also the current responsible for establishing the necessary magnetic field.

Inspection of Fig. 4-1 allows a loop voltage equation to be written for the circuit shown.

$$V_T = E_a + I\ (R_f + R_a) \tag{4.1}$$

$$E_a = V_T - I\ (R_f + R_a) \tag{4.2}$$

but

$$E_a = \frac{P\Phi Z n}{60a} \times 10^{-8} \tag{4.2}$$

(from Eq. 2.10)

or

$$E_a = K\Phi n$$

Then if the expression for E is substituted into Eq. 4.1, a new expression for motor speed is obtained.

$$n = \frac{V_1 - I_a (R_f + R_a)}{K\Phi} \qquad (4.3)$$

The speed equation allows a number of deductions to be made about the speed of the motor. Since for a constant torque load I_a and Φ remain constant, it shows that speed can be varied by either varying the applied voltage or increasing the resistance of the motor circuit. Conversely, if the applied voltage is held constant and load torque is increased, both I_a and Φ will also increase. It can easily be seen that the net result is that speed will decrease with increasing load.

The negative slope of the torque-speed curve of the series field motor is shown in Fig. 4-2. The motor is characterized by a very high starting torque and a high no-load speed. With some machines it may be necessary to take precautions to limit the light-load speed. Otherwise, motor speed might become so great that the centripetal stresses on the rotor would be great enough to tear it apart.

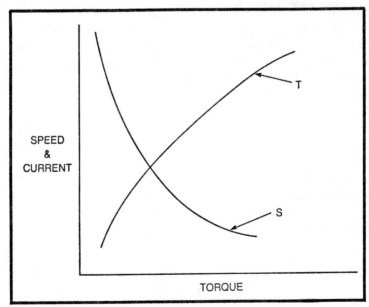

Fig. 4-2. Speed-torque characteristics of the series motor.

The series-field motor has a torque-speed characteristic that makes it a very good choice for a number of diverse applications. In small sizes it is used to power portable and cordless appliances and hand tools.

At the opposite end of the spectrum of application is the traction motor. In this use the high starting torque of the series motor is exploited to provide high acceleration capability. Traction motors, which will be treated in greater detail in a later section, range in size from fractional horsepower up to very large integral horsepower sizes.

The field winding of a series connected motor is typically composed of a relatively small number of turns. The wire size will usually be relatively large also, approximately the size of the armature conductors. This is because the wire must be large enough to handle large armature currents without overheating. But since the current is high, a large number of turns is not necessary to produce the ampere-turns needed for the field strength.

A point of interest is the fact that the series motor will run with either an ac or dc voltage exciting it. This is possible because the same current flows through the field and armature circuits. Then as the current flow changes, as with an ac excitation, the developed force remains in the same direction.

THE SHUNT-FIELD MOTOR

The shunt-field motor is one in which the field winding is connected in parallel with the armature winding. This connection is shown schematically in Fig. 4-3. The shunted field winding can be excited from the same voltage source as the armature (self-excited) or it can be excited by means of an independent voltage source (separately excited). In either case it can be seen that the field winding is excited by a constant voltage, and the current it carries is not affected by the armature generated voltage. The shunt field motor is thus seen to be a constant-flux machine, i.e., for a given value of applied voltage, the flux does not change as load torque and speed may vary.

By inspection, a loop voltage equation can be written for the shunt motor. If this is done we obtain

$$V_T = E_a + I_a R_a \qquad (4.4)$$

$$E_a = V_T - I_a R_a \qquad (4.5)$$

$$n = \frac{V_1 - I_a R_a}{K\Phi} \qquad (4.6)$$

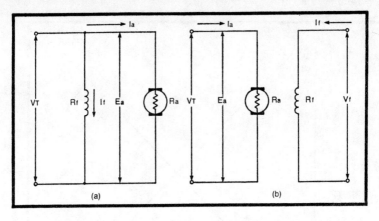

Fig. 4-3. Shunt field motors, (a) self-excited connection and (b) separately excited connection.

It is seen that Eq. 4.6 is similar in form to Eq. 4.3 for the series motor. However, in the case of the shunt field motor the magnetic field strength is a constant. If the magnetic field strength is a constant, then the developed torque is directly proportional to the armature current. This relationship can be written as

$$T = K_1 I_a$$

or

$$I_a = \frac{T}{K_1} \tag{4.7}$$

If Eq. 4.7 is substituted into Eq. 4.6 an expression is obtained relating the motor speed to torque,

$$n = \frac{V_1 - \dfrac{TR}{K_1}}{K\Phi}$$

or

$$n = \frac{V_1}{K\Phi} - \frac{R_a T}{K_1 K\Phi} \tag{4.8}$$

Equation 4.8 is recognized as being the characteristic of a straight line curve. The no-load speed can be calculated by setting T equal to zero, which reduces the equation to,

$$N_o = \frac{V_1}{K\Phi} \tag{4.9}$$

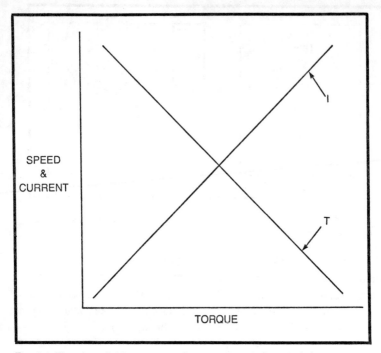

Fig. 4-4. The shunt-field motor speed-torque-current characteristic.

The motor's stalled torque can be calculated by setting n equal to zero and solving for T. If this is done, the expression for stalled torque becomes

$$T_S = \frac{V_1 K_1}{R_A} \qquad (4.10)$$

The speed-torque-current characteristic of a shunt-field motor is shown in **Fig.** 4-4. The linear characteristics of this type of motor make it very easy to control. Speed may be varied by changing the magnitude of the applied voltage. It can also be controlled by varying the field voltage which has the effect of changing the air-gap flux, Φ. Because of these characteristics, which are so very compatible with controlled speed drives, the shunt-field motor is extensively used in adjustable-speed drive applications. In such an application a controller is used to sense motor speed and automatically adjust the voltage applied to the motor, either upward or downward, as required. The motor speed responds by being proportional to the applied voltage.

The field winding of a shunt-connected motor, while functionally similar to that of the series-field motor, is very much different in its practical aspect. It will usually consist of many turns of a relatively fine wire. This high resistance winding is necessary to hold the field current to a level which does not cause it to overheat. Because of the practical differences in their field windings, a series-field motor cannot be arbitrarily reconnected as a shunt field motor or vice versa. Shunt field motors come in all sizes from small subfractional hp up to very large integral hp sizes.

THE COMPOUND-FIELD MOTOR

The compound connection is shown schematically in Fig. 4-5. As the name suggests, this motor combines the features of a shunt and a series motor by incorporating both windings. The field windings are usually connected so their associated magnetic fields act in the same direction. This type of winding connection is called cumulative compounding. If the windings are connected so their magnetic effects are opposite it is called a differentially compounded connection. Differential compounding of motor fields will often lead to speed instability and is usually avoided.

The speed-torque characteristic of a compound field motor can take on a wide range of features, depending upon the relative strengths of the shunt and series fields (Fig. 4-6). If the shunt-field winding is relatively weak, the speed-torque curve will resemble

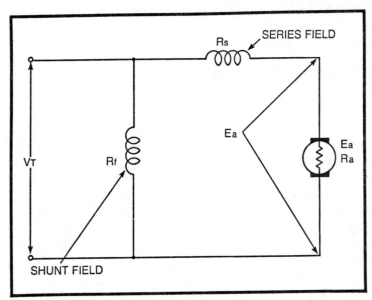

Fig. 4-5. The compound-field connections.

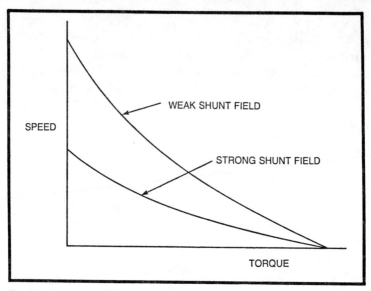

Fig. 4-6. Compound-field motor performance characteristics.

that of the series motor. On the other hand, if a strong shunt field is used, the motor curve will approach the straight line of the shunt-field motor.

Compound-field motors are used in a wide variety of applications. The strong shunt type is suitable for use in adjustable speed drives. Weak shunt-field motors (strong series-field) find application in traction motor drives.

PERMANENT-MAGNET MOTORS

The permanent-magnet motor (Figs. 4-7 and 4-8) is one in which permanent magnets are used to produce the necessary field in place of the electromagnets of wound-field motors. Because its field is established by a permanent magnet, the air-gap flux is constant and the resulting torque-speed characteristic is similar to that of a shunt-field motor.

Permanent magnet motors are an increasingly important commercial product. This is especially true since the advent of ceramic magnets has allowed the design of motors of competitive performance at manufacturing costs well under that of a comparably rated wound-field motor.

PM motors are usually designed with ceramic magnets, although not exclusively so. The ceramic magnets are an order of mangitude less in cost as compared to the high energy alnico magnets, and also have a very high coercivity. This feature makes them resistant to demagnetization and reduces the size of the magnet required. On the other hand, the alnico family of magnets provide very high flux densities with resulting high motor performance but are susceptible to demagnetization. Because of their high cost and other limitations the alnico's are used primarily in high performance servomotors.

PM motors have been built for a very wide range of applications, ranging from simple toy motors to sophisticated space and

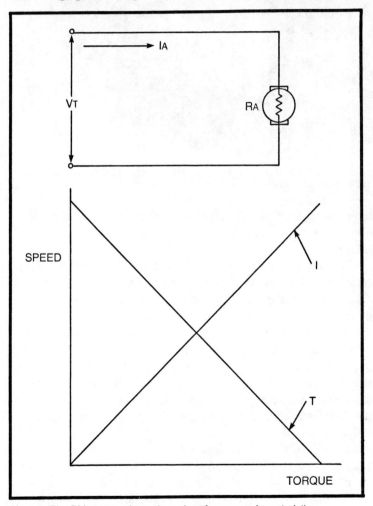

Fig. 4-7. The PM motor schematic and performance characteristics.

Fig. 4-8. The stator of a permanent-magnet dc motor. (Courtesy Gould Inc., Electric Motor Division.)

computer applications. The range of horsepower ratings that have been successfully built goes from subfractional horsepower up into small integral horsepower sizes.

The PM motor does possess a number of disadvantages when compared to a wound-field motor. Since the air-gap flux is fixed by the geometry of the magnet and the air gap, this parameter cannot be varied to achieve speed control. In this regard, a wound-field motor is more versatile, since it does allow the field strength to be varied. A second disadvantage is that a PM motor can be demagnetized by a high enough pulse of armature current. This is not likely to happen since most motor controllers do have inherent current limitations. Nevertheless, demagnetization of the field magnets is a physical possibility and must be guarded against. Finally, because PM motors are designed to minimize manufacturing costs, they are usually built without interpoles, which improve commutation. Because of the lack of interpoles, especially in ratings from ¾ hp and up, the PM motor has a tendency for visible sparking under the brush. This brush sparking is often taken as an indicator of poor brush-life expectancy. This is not always a good indicator since other factors, such as current density and heat generation at the brush, may be predominant factors in brush wear.

The PM motor does possess a number of distinct advantages over its wound-field counterpart. The most important advantage of course is the manufacturing cost economics that are made possible. When a motor is designed to optimize the possibilities of a permanent-magnet field, the result is different than what it would be for a wound-field motor. The overall effect, though, is to significantly reduce the material and labor costs of the PM motor. Whereas with a wound-field motor, increasing the number of poles also increases costs, the opposite is true with the PM motor. By using a larger number of field magnets, the PM motor achieves overall material economics. This effect is seen in Fig. 4-9. The PM motor will have a larger diameter rotor with enlarged slots and wide slot openings. The volume required for the field magnets is greatly reduced from that of its wound field counterpart.

Another advantage of the PM motor is the fact that the magnetic field necessary for the motor action is provided without the necessity of an external power supply. This means that the controller is relieved of the need to provide a field winding voltage, with a resulting cost savings.

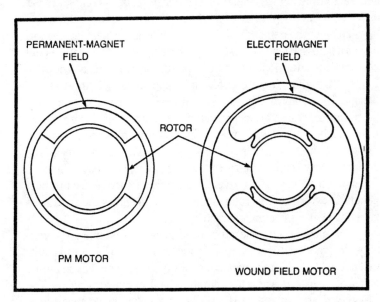

Fig. 4-9. Relative sizes of wound-field and PM motors of the same horsepower rating.

A third advantage of the PM motor is the reliability factor, which it does provide. The magnetic field is always on. Its function can not be affected by a failure of the field voltage supply or the field winding itself. The magnetic field does provide a slight detent torque to hold an angular position sufficient in some cases to eliminate the need for a holding brake.

SERVOMOTORS

Because of its outstanding controllability features, the dc motor is a natural for use in closed loop control systems. When used in combination with a precision tachometer generator and a high-gain amplifier, it is possible to achieve speed regulation of ½% over a speed range ratio of 1000 to 1.

In order to be useful in a high performance control system, it is necessary to have a motor with highly developed special features. The motor must have the ability to instantly accelerate its load and follow closely a command signal. The inductance, or energy storage feature, of the motor must be kept low so as to minimize time delay between command signal and shaft actuation. Finally, the motor should also have the ability to readily dissipate joule heat which is generated from high acceleration and deceleration currents often at high cyclic rates.

To meet the demands of the high performance system, the low inertia ironless armature design was developed. This armature concept has been developed along two parallel approaches as shown in Fig. 4-10. Although quite different in appearance, both armature designs show similar advantages and performance characteristics. Since the armatures are constructed without any moving iron, they have an inherently low moment of inertia. For the same reason the inductance of the winding is also very low. These traits make the armature very responsive to voltage signals. It can readily be seen that both armatures have a very high ratio of surface area to heat generating volume. This ratio of area to volume gives the armatures superior heat dissipating characteristics.

The motors and generators used in servo systems are invariably designed with permanent-magnet fields. The permanent-magnet field gives the advantage of always "being on," even during quiescent periods, without generating heat in a field winding. The PM motor armature also is usually of a lower inductance which improves its responsiveness.

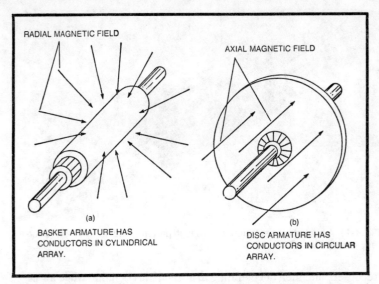

RADIAL MAGNETIC FIELD

AXIAL MAGNETIC FIELD

(a)

(b)

BASKET ARMATURE HAS
CONDUCTORS IN CYLINDRICAL
ARRAY.

DISC ARMATURE HAS
CONDUCTORS IN CIRCULAR
ARRAY.

Fig. 4-10. Two examples of servomotor development to reduce inertia and improve motor response, (a) the basket armature and (b) the disc armature.

Servomotors are built with either ceramic magnets or alnico magnets. The ceramic magnet design is by far the more common because of lower manufacturing costs associated with it. The ceramic magnets are not only much lower in cost themselves when compared to alnico magnets, but also allow other cost economies in the manufacture of the motor. On the other hand, the alnico magnets are capable of supporting much stronger magnetic fields. For example, it is possible to concentrate an air-gap flux density of 10,000 gauss using alnico 5-7 magnets, while the best ceramic class is capable of 3500 gauss. Since motor torque is directly proportional to the air-gap flux per pole, it can easily be seen that the highest performance motor will utilize alnico magnets.

The highest performance servomotors manufactured are those supplied to the computer peripheral equipment industry for use with digital magnetic tape drives. In this application the motor directly drives the capstan, which moves the magnetic tape over the read/write head. The trend in the computer industry has been, over the years, to denser recording on the tape. This means that the blank space between recorded bits of information has been made smaller and smaller. In terms of motor characteristics, this requirement means the motor must be capable of accelerating its load to a speed of 2000 rpm, running at a closely controlled speed while data is being

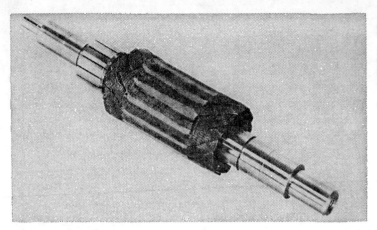

Fig. 4-11. A basket type armature from a high performance servomotor. (Courtesy Electro-Craft Corp.)

recorded, and then decelerating to a standstill condition while awaiting the next command. The period for the entire cycle will be about 0.005 seconds and the duty may be continuous. In other words the motor is turned on and off several hundred times a second.

If a conventional design motor, i.e., with iron core, were put into an application as described in the preceding paragraph, there would be no response to the switching voltage other than a vibration. During this time, the armature would overheat to damaging temperature. The large energy storage associated with a high inductance and high inertia armature would prevent useful rotational movement.

The armature structure of a high performance servomotor is shown in Fig. 4-11. The armature is built without a steel core. The conductors are formed into a hollow cylinder held in rigid form by a high strength epoxy resin. Great pains must be taken to keep the inertia as low as possible in all parts of the rotor structure. For that reason the armature conductors will be aluminum rather than copper. The commutator will be made with as small a diameter as possible. In some applications, the inertia becomes so critical that the shaft may be made from a material such as titanium which combines a high modulus of elasticity with a low specific gravity.

Servomotors are not rated in terms of horsepower output as is the case for conventional motors. Rather the important parameters are the torque constant (developed torque per amp of input), the armature resistance, and the rotor inertia. Since the servomotor is usually starting and stopping at high cyclic rates, its steady-state output rating is of less importance than its performance in a transient mode. Thus from the significant servomotor data that is provided, the servo engineer can calculate the frequency response of the motor and the power dissipation for a given inertial load and duty cycle.

TACHOMETER GENERATORS

A tachometer generator is a machine used as a generator where the voltage is used as a signal rather than as a power source. In its use it can be contrasted to a power generator or an exciter generator whose purpose is to supply electric energy. When intended as a power source a generator is carefully designed to maintain a constant voltage even though the speed may vary. The tachometer generator, or tach generator as it is commonly called, is intended to provide an output that varies directly with shaft speed. Such a device is very useful in providing a measure of shaft speed and in a form that can be fed back to a motor input to modulate it and thus regulate a speed. The block diagram for such a speed control system is shown in Fig. 4-12.

This figure shows how a tach generator is used to close the loop in a speed control system. Its output voltage subtracts from that of a reference input at the summation point. The system works in this manner: The input voltage that corresponds to the desired shaft

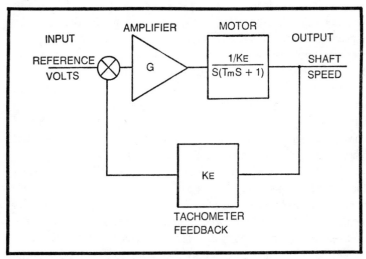

Fig. 4-12. The tach generator in a controlled speed system.

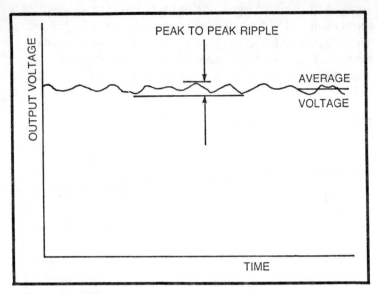

Fig. 4-13. The ripple voltage in the output of a dc generator.

speed is set on the control panel. The motor responds to the voltage and runs at the desired speed. If the load on the output increases, the motor speed tends to fall. The tach generator is coupled to the same shaft as the motor, so as the speed decreases the output voltage of the generator also decreases. With a decreased feedback voltage being subtracted from it, a larger error voltage is fed into the amplifier and a larger voltage is applied to the motor. With a larger applied terminal voltage the motor will tend to speed up. This process continues until an equilibrium is reached. In this way it is possible to control motor speed within very close limits.

Because of the nature of its application, a tach generator does not have to be large or have significant power capacity. It is important, however, for the generator to have a low ripple on its output voltage. It should also have a linear voltage-speed characteristic. These two characteristics are shown in Figs. 4-13 and 4-14.

The ripple in the output of the tach generator is due to the fact that the output is actually made up of a number of sinusoidal components. The number of sinusoids is determined by the number of armature coils. As the number of coils is increased the ripple decreases until it approaches the ideal, flat characteristic. Ripple is usually measured and specified by taking the total excursion of the voltage, from peak to peak, and expressing it as a percentage of the average output. For example, a tach generator with an output of 10 volts at 1000 rpm and rated with a 3% peak-to-peak ripple means that at a constant 1000 rpm the voltage will be fluctuating between 10.15 and 9.85 volts.

The importance of the ripple characteristic is due to its effect on the speed control system. This effect can be understood by referring again to Fig. 4-12. In order to keep the system control high it is necessary to use a high gain amplifier. The amplifier detects the voltage variations due to generator ripple and transmits them into the voltage applied to the motor. If the motor is a highly responsive servomotor it will oscillate in speed with the ripple and increase the error of the system. In high performance systems it is necessary to limit the amount of voltage ripple. It is also important that the machine have a linear voltage-speed characteristic, as shown in Fig. 4-14. In order to achieve good linearity, a tach generator must have a constant magnetic field strength so that the voltage at a given value of rpm is always the same, and so the generated voltage will be in direct proportion to the speed. For this reason PM fields are ideal for tach generators. A second factor in achieving good linearity of output voltage is the ratio of the armature impedance to its load impedance. This ratio must be kept very small to avoid a self-regulating effect at high speeds (and high voltages). This effect is shown in Fig. 4-15. In order to avoid this type of nonlinearity the armature resistance should be kept small in comparison to the summing resistor. Or

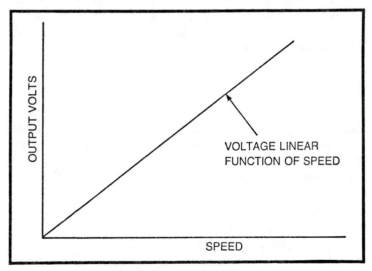

Fig. 4-14. Voltage-speed characteristics of a tach generator.

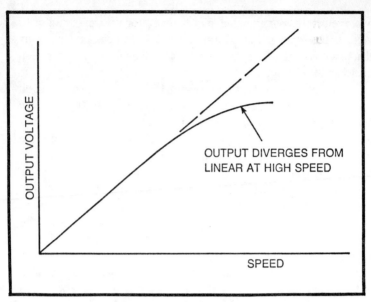

OUTPUT VOLTAGE

OUTPUT DIVERGES FROM
LINEAR AT HIGH SPEED

SPEED

Fig. 4-15. Nonlinearity of a tach generator output due to the regulating effect of the armature winding impedance.

conversely, the summing resistor should be at least an order of magnitude greater than the armature resistance.

The best type of tach generators are designed with alnico-magnet fields. The high magnetic flux of the alnico provides a large output voltage gradient. The alnico magnets also have a very small temperature coefficient associated with their flux. So a tachometer generator with an alnico-magnet field will be relatively little affected by ambient temperatures.

BRUSHLESS MOTORS

Over the years the controllability of the dc motor has always made it a popular choice for adjustable speed drives and for closed loop servo systems. Similarly the high starting torque and positive damping of the torque-speed characteristic have made it very appropriate for traction applications. Despite the obviously desirable features that are offered, however, the dc motor has a negative feature which has confounded several generations of engineers. The main drawback of the dc machine is due to the conventional manner of commutating the coil currents, viz, the use of sliding contacts (or brushed) with high wear rate and limited life associated with them.

The brush wear that accompanies many applications requires that rigorous maintenance schedules be adhered to. If brushes are not replaced at the scheduled interval or if the brush wears away at a faster rate than anticipated, the brush shunt can be pushed into the commutator, resulting in severe damage and necessary repair costs. As a consequence of these shortcomings of conventional commutators and brushes, motor engineers have long been intrigued with the idea of developing a machine with dc motor characteristics but without brushes and their attendant problems.

In the past two decades technological developments in other fields have resulted in components which, when applied to a unique machine design, have made the brushless machine a reality. These technological developments include the barium ferrite ceramic magnets and solid state switching devices such as silicon controlled rectifiers (SCRs) and power transistors. The magnets are capable of supplying a strong magnetic field while attached to a rotating member. The switching devices can be gated to, upon command, either allowing current flow or interrupting it.

This is the way most brushless dc motors work. Refer to Fig. 4-16. The figure shows two of the three main parts of a brushless

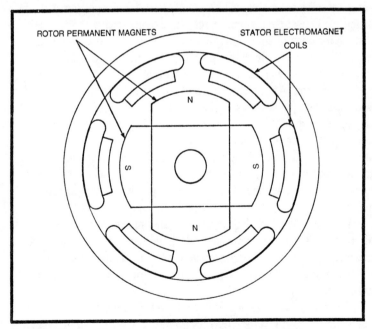

Fig. 4-16. The brushless dc motor with permanent-magnet rotor.

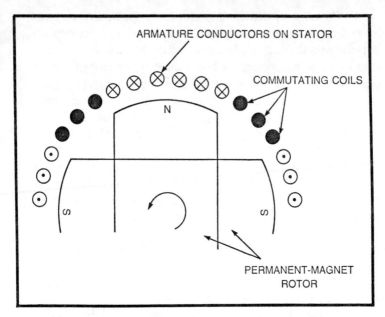

Fig. 4-17. Magnet polarity and current flow in a brushless dc motor.

motor. A rotor assembly will mount an even number of permanent magnets. These permanent magnets provide the magnetic field medium for energy conversion in the same manner that the wound field or PM field of conventional designs function. The stator of the brushless motor will very closely resemble that of a polyphase induction motor. As the rotor is made to rotate, the magnetic field associated with its magnets will also move with respect to the stator. The result is a $d\Phi/dt$ effect in each winding group as the magnet moves past it. As we have developed in a previous chapter, a time varying magnetic flux will induce a voltage in a coil which encircles it. As the rotor is caused to turn, the stator windings will experience a generated voltage whose magnitude is proportional to the rotational speed.

We have learned that generator and motor effects are two different views of the same natural phenomena. So it is with the brushless motor; if a voltage is applied to the armature coils mounted on the stator the necessary conditions exist for torque development and motor action.

The only remaining problem for the motor designer is to devise a scheme for exciting the appropriate stator windings. For just as was the case with the conventional machine (Fig. 2-5) it is necessary for all of the stator conductors opposite a particular magnet polarity to have currents flowing in the same direction. Also, the current direction must change as the polarity of the magnet opposite it changes. Figure 4-17 shows all of the currents in the slots opposite the north pole to be into the paper. The currents in the slots opposite the south pole flow out of the paper. The currents in the slots between poles are in the process of commutating, changing from a north pole direction of flow to a south pole flow.

The switching is done by the solid state devices, SCRs or transistors (Fig. 4-18). The switching devices are triggered on or off by some type of shaft angular position sensor, which is the third essential portion of the machine. The most commonly used techniques for triggering the armature circuit switches are optical encoders and Hall-effect probes. The function of the position sensor is the same regardless of the type used. It senses the angular position of the rotor magnet and triggers the appropriate switches so that stator windings in the proper slots are excited and develop torque.

The torque-speed characteristic of a brushless motor will be similar to that of a conventional permanent-magnet motor or shunt-field motor (Fig. 4-19). The torque-speed curve is linear with negative slope and torque is proportional to current.

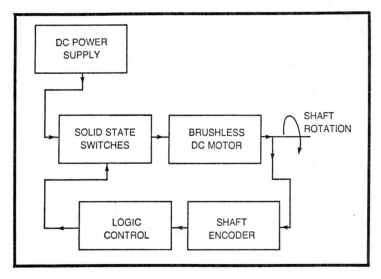

Fig. 4-18. Shaft position encoder used to switch windings in a brushless dc motor.

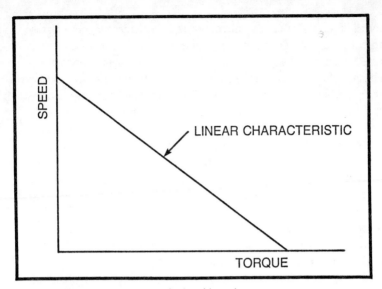

Fig. 4-19. Motor characteristics of a brushless dc motor.

Brushless motors have been developed as a commercial item in very small subfractional horsepower ratings. Designs have also been developed in fractional horsepower sizes and small integral horsepower sizes. But the use of the larger motors is not widespread. The present state of the art is such that the cost of position sensor and commutation circuits more than offsets the cost of a conventional mechanical commutator and brushes. As a result the present offerings of brushless motors carry a substantial premium price over a conventional motor of equal rating and are limited in application.

UNIPOLAR MACHINES

The unipolar machine, sometimes called homopolar machine, is an anomaly among dc machines. While all other machines are characterized by bipolar magnetic fields, the unipolar machine has a single magnetic polarity around its entire periphery (Fig. 4-20).

Although just two poles are shown for the machine in Fig. 4-20(a), it should be remembered that a bipolar machine can have a number of poles that is equal to any multiple of two. On the other hand, it is seen by referring to Fig. 4-20(b) that the unipolar machine has an air-gap flux which is directed inward around its entire periphery. A side view of such a unipolar design is shown in Fig. 4-21. The unipolar machine has a number of features which are very desirable and also features which limit its practical usage.

In Fig. 4-21, it is seen that current is introduced at one end of the armature. It passes through the length of the conductor and out the opposite end. The armature of a unipolar machine consists of but a single conductor. It can be made in the form of either a copper cylinder or disc. This unique feature has tremendous implication on the voltage characteristic of the machine. If the voltage Eq. 2.10a is again referred to

$$E = \frac{PZ\Phi n}{60a} \times 10^{-8}$$

it is seen that with but a single pole P and a single conductor Z, the only possibility for obtaining a practical level of generated voltage is to have a very high value of air-gap flux Φ and to operate at a high speed n. So it is that unipolar machines in actual use are of a very large physical size to provide the large magnetic flux required for voltage generation.

In a unipolar machine there is no commutation of the armature current. Therefore, the generated voltage is absolutely constant at constant speed without any of the ripple voltage typical of commutator machines.

The unipolar machine does posses a physical robustness and a tremendous current capacity. In order to handle the large currents that must be collected from the armature, liquid metal is most often

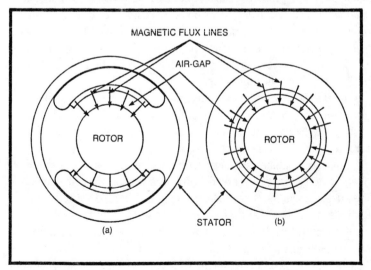

Fig. 4-20. The magnetic field orientations of bipolar and unipolar machines, (a) the bipolar machine and (b) the unipolar machine.

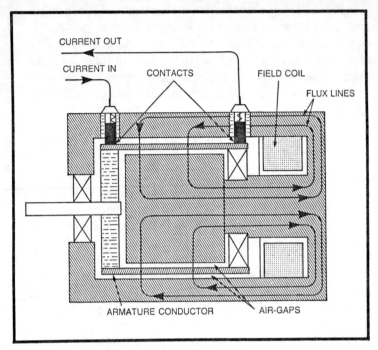

Fig. 4-21. Side view of field in a unipolar machine.

used in place of a graphite brush. A material such as liquid sodium is contained in a labyrinth at either end of the armature. The armature protrudes into the liquid metal and current is passed through the liquid metal and into the armature. This type of current collector has a much lower contact drop than what would be obtained with conventional types of graphite brushes.

A unipolar machine can be designed with either a permanent-magnet or an electro-magnet field. The new rare-earth magnets with very high coercivity and high flux density have very desirable characteristics for unipolar applications.

Unipolar machines are very useful where large amounts of low voltage electric energy are used. For this reason they are sometimes used in aluminum reduction plants where an electric load of hundreds of thousands of amperes at several volts is ideal for a unipolar generator. Motors built with a unipolar design are inherently of high efficiency and high heat dissipating characteristics. These features make it possible to achieve startling economies with motor materials. A unipolar machine might be less than half the size of a similarly rate machine of conventional design.

94

Chapter 5

Operating Characteristics of DC Machines

In Chapter 2 a number of fundamental laws were used to develop working equations. The equations can be used to calculate generated voltages and machine torques. The accuracy of the results obtained by use of these equations can vary greatly depending upon how skillfully they are applied. It certainly would not be unusual for the novice to attempt to apply the equations to an actual machine and find his calculations greatly different from test results. Despite this rather common occurence it should be kept in mind that electromagnetic theory has withstood the test of time and when the calculations are carefully carried out, they will agree closely with the test results. Invariably, when a large discrepancy does occur between calculation and test, careful analysis will reveal that some operating condition of the machine departed significantly from the assumptions on which calculations were based. Thereupon, if a proper allowance is made for the particular operating condition, perhaps a temperature effect, good correlation will be obtained.

The sections of this chapter discuss various parameters and characteristics which are used to rate, specify, or otherwise describe various aspects of machine performance. A good understanding of these characteristics is invaluable to successful application of dc machines.

MACHINE RATINGS

The most commonly used rating system for dc machinery is to specify and output power at a certain speed and terminal voltage condition. In the case of a motor the unit of output power is the horsepower. While in the case of a generator the unit of output is the watt or kilowatt.

Over the years an effort has been made to standardize on machine speeds. As a consequence the standard speeds in effect today are: 1150, 1725, 2500, and 3500 rpm. Of course a dc machine, in contrast to an induction motor, can easily be designed to run at almost any speed. However, if a standard design is desired it will have to be selected at one of the standard speeds.

Similarly, standard voltages have been developed. The most commonly encountered voltage for fractional horsepower (FHP) machines is 90 volts. The 90 volt rating is consistent for use with rectified 115 volt, 60 hertz supply. As machines get into integral horsepower (IHP) sizes, the standard voltage is 180 volts, selected for its compatibility with rectified 230 volt, 60 hertz power.

In addition to the voltage standards which have developed for use with 60 hertz rectified ac power a large number of dc motors are run with battery power. Voltages commonly encountered for use with batteries range from 12 volts up to 48 volts. The low voltage ratings are, of necessity, limited to fractional horsepower motors. However, 48 volt designs have been successfully built in power ratings up to 30 hp. The problem posed by the current requirement for a 48 volt, 30 hp motor can be appreciated by a simple calculation. Assuming an 85% efficiency, the rated current for such a machine will be,

$$I = \frac{0.30 \times 746}{0.85 \times 48} = 549 \text{ amps}$$

Such a current level requires extremely large armature conductors to keep the current density at a tolerable level. In addition, very large sized brushes and commutator are required for the function of transferring electric current to the armature.

A machine characteristic that is getting an increasing amount of attention is operating efficiency. Efficiency is the ratio of output power to input power expressed as a percentage of the input power. Or,

$$\eta = \frac{\text{Power out}}{\text{Power in}} \times 100\%$$

The ever increasing cost of energy will provide an additional impetus to obtain higher machine efficiencies so as to reduce their operating costs. Typical machine efficiencies range from the low 70% for FHP sizes up to 90% for small IHP sizes. In addition to the extra cost for electric energy, low efficiency motors pose a special problem when used with portable battery supplies. In such cases the low efficiency motor can significantly reduce the time available between recharging cycles for the battery.

WINDING IMPEDANCES

The user of dc machines usually is not interested in the impedance characteristics unless they directly affect the application. Because of the increasing use of rectified ac power and chopper controllers with battery power, the complex impedance of dc machine windings has taken on added significance.

The various coil windings in a dc machine constitute electronic circuits in much the same manner as any other assembly of circuit elements. As such, the machine circuits have the usual parameters of inductance, capacitance, and resistance. In most practical machines the capacitance is a negligible effect as compared to the inductance and can be neglected.

Figure 5-1 shows schematically the windings in a compound-field motor. A similar type of circuit can be drawn for any other machine type. Showing the circuit elements as they are implies that a lumped parameter approach can be used in machine analysis. Such is

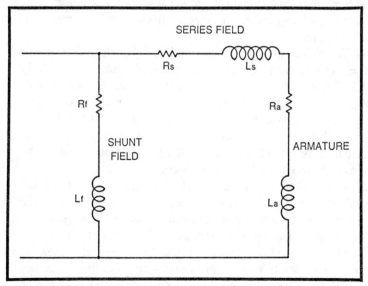

Fig. 5-1. The circuit schematic for a compound-field machine.

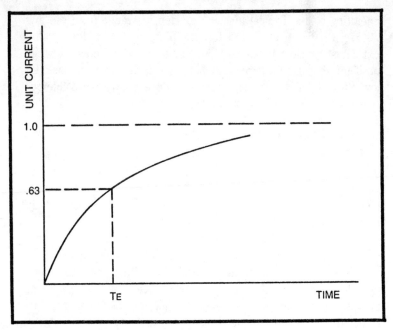

Fig. 5-2. The current response in an RL circuit.

the case and later chapter will develop an equivalent circuit analysis which can be used to calculate machine performance. For now it is sufficient to say that circuit elements that characterize power dissipation (resistance) and energy storage (inductance) are present in the dc machine.

The circuit elements as shown in Fig. 5-1 are largely responsible in determining a machine's limitations. If a voltage is suddenly applied to the circuit shown, the current response will be expoential (Fig. 5-2). Its shape will be determined by the circuit electrical time constant, the ratio of the winding inductance to its resistance. It can be seen that since instantaneous torque is proportional to the instantaneous current level, a motor's ability to accelerate a load can be profoundly influenced by its inductance.

It can also be seen that if a dc voltage is applied to the circuit, the resulting current will increase from zero until, at a point in time, the current will reach an ultimate value limited only by circuit resistance. This circuit resistance is proportional to the watts of input power that are converted into heat. The heat so generated causes the armature temperature to rise to its limiting temperature. It can then be seen that there is a definite limit on how much power can be dissipated in heat before damaging temperatures occur in a winding. The winding resistance thus determines the maximum current that can be fed into a winding.

TEMPERATURE EFFECTS DUE TO RESISTANCE

All machine ratings are based upon a particular temperature condition existing in the machine winding. The rated condition is also based upon a standard ambient temperature condition of 40° centigrade. If a machine is used in an application where the ambient differs significantly from the 40°C standard, it is possible for the machine characteristics to also vary significantly. In the extreme case it may be necessary to derate the machine so as to avoid excessive temperature rise.

The resistance of an electrical conductor is influenced by its temperature. In general all metal conductors have a positive temperature coefficient of resistivity, their resistance increasing with temperature. It is accepted practice to specify winding resistances at a standard (room ambient) temperature of 25° C. An expression that relates resistance at some other temperature to its standard value is given by

$$R' = R \; [1 + \alpha(\Delta T)] \qquad (5.1)$$

In Eq. 5.1, R is the value of resistance at 25° C, R' is the resistance at some other known temperature, α is the temperature coefficient of resistivity, and ΔT is the change in temperature using 25° C as the reference. Equation 5.1 provides a means for calculating values of resistance at elevated temperatures. It also provides a means of calculating a temperature rise by making more convenient measurements of resistance at 25° C and again at the operating condition of interest. Table 5-1 shows some typical coefficients.

Example. A motor with copper windings has an armature resistance of 0.5 ohm at the standard temperature of 25° C. After the motor operates a sufficiently long time to reach thermal equilibrium, the resistance is measured as 0.7 ohm. What is the average temperature rise of the armature?

The temperature rise is calculated by rearranging the terms of Eq. 5.1 and solving for ΔT, so

$$\Delta T = \frac{(R'/R) - 1}{\alpha}$$

Table 5-1. Temperature Coefficients of Resistivity for Typical Conductors

Conductor material	Temperature coefficient of resistivity (α)
aluminum	0.00408
copper	0.00393
platinum	0.00392
silver	.0.0041

then

$$\Delta T = \frac{(0.7/0.5) - 1}{0.00393}$$

$$\Delta T = 102° \text{ C}$$

TEMPERATURE EFFECTS ON TORQUE-SPEED CHARACTERISTICS OF A SHUNT-FIELD MOTOR

Referring to Fig. 5-1, it can be seen that if the circuit elements are subjected to a temperature increase, the resistances will also increase as predicted by Eq. 5.1. The increase in resistance, which is typically on the order of 40% for a class F machine, is sufficient to alter the torque-speed characteristics of the machine. This effect is particularly true for the shunt-field motor.

If a motor is taken without a prior warm-up, the field current Ip is determined by the cold value of the field winding resistance. The field current in turn establishes the air-gap flux, Φ, of Eq. 2.10. As the field winding resistance increases due to thermal warm-up, field current decreases in inverse proportion. The decreased value of field current causes a reduction in magnetic field intensity in the air gap. The result is that the machine will tend to run faster.

The temperature rise effects on the torque-speed characteristic of a shunt-field motor are shown graphically in Fig. 5-3. If an application requires that a machine run precisely at a preset speed, it may be necessary to design around the temperature rise effect, so as to minimize speed change.

VOLTAGE CONSTANT OF A MACHINE

A very useful expression can be obtained to provide insight into the behavior of shunt-field and PM machines. Consider the voltage equation derived in Chapter 2, Eq. 2.10.

$$E = \frac{PZ\Phi n}{60a} \times 10^{-8}$$

Once a machine is designed, the factors P, Z, and a are fixed and become a constant value. In a shunt-field machine and also in a PM machine the air-gap flux is also a fixed value. The Eq. 2.10 can be rewritten:

$$E = k_E n \qquad (5.2)$$

where
$$k_E = \frac{PZ\Phi}{60a} \times 10^{-8} \qquad (5.3)$$

The constant k_E has the dimensions of volts/rpm. Equation 5.2 can be used to make calculations directly and it will also be a very useful form for subsequent analytical developments.

Example. A shunt-field machine has the following design features:

$P = 2$ poles
$Z = 1200$ armature conductors
$\Phi = 464{,}170$ lines
$a = 2$

What is the voltage constant of the machine and what voltage is generated at 1800 rpm?

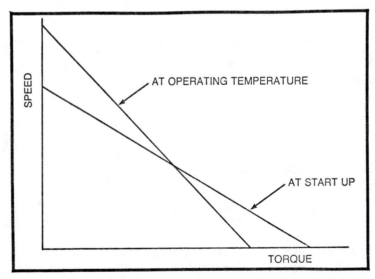

Fig. 5-3. The effects of heating and temperature rise on the characteristics of the shunt-field machine.

Use Eq. 5.3 to calculate k_E

$$k_E = \frac{2 \times 464{,}170 \times 1200 \times 10^{-8}}{60 \times 2}$$

$k_E = 0.0928$ [volts/rpm]

Use Eq. 5.2 to calculate the generated voltage,

$$E = 0.0928 = 1800 \times 167 \text{ volts}$$

The voltage constant of a machine is applicable to motors or generators. In either case it is a measure of the voltage generated in the armature conductors for each unit of rotational speed. If the IR drop in the armature is negligible as it is at a no-load condition, the voltage constant can be used to calculate motor speed.

Example. Calculate the no-load speed of a machine with a voltage constant k_E of 0.0678 [volts/rpm] when it is excited with 180 volts.

Use Eq. 5.2 after solving it for n,

$$n = \frac{180}{0.0678} \quad = 2655 \text{ rpm}$$

TORQUE CONSTANT OF A MACHINE

Torque-constant is also a parameter that is applicable to shunt-field and PM motors. It is obtained by considering the expression for torque, Eq. 2.14, with an added constant.

$$T = \frac{k_1 PZ\Phi I}{a}$$

(Constant k_1 will depend upon the dimensions of the torque units used.) As was the case with voltage constant, once a design is established all of the factors in the torque equation are fixed and the equation can be rewritten as

$$T = k_1 I \tag{5.4}$$

where (with ϕ in maxwells)

$$k_T = \frac{k_1 \, PZ\phi \times 10^8}{a} \tag{5.5}$$

The dimensions of the torque constant can be lb-ft per amp or oz-inch per amp or any other consistent units of torque so long as the value of k_1 is properly chosen.

The torque constant of a machine can be very useful. Again, as was the case with the voltage constant, the concept of a torque constant is applicable to a machine used in either a generator or a motor mode of operation. It is an indication of the electromagnetic torque that is developed for each ampere of electric current flowing through an armature winding. If applied to a motor it will provide a means of calculating the output torque of the machine. When applied to a generator, it provides a means of calculating the torque loading upon the prime mover.

Example. A shunt field machine has the following design features:

P = 2 poles
Z = 600 armature conductors
Ω = 925,000 lines
a = 2

a. What is the torque-constant of the machine?
b. If the machine is used as a generator and loaded so that a 10 amp current is flowing, what electromagnetic torque is developed?

from Eq. 5.5,

$$k_T = \frac{(0.117 \times 10^{-8})2\,(600)\,925,000}{2} = 0.649 \text{ [lb-ft/amp]}$$

use Eq. 5.4 to calculate the electromagnetic torque

$$T = 0.694 \times 10 = 6.49 \text{ lb-ft}$$

The concept of torque and voltage constants can also be applied to machines other than the shunt-field and PM types. However, the concept is not nearly so useful or conveneint to use in other types of machines. This is because the magnetic field strength of the shunt-field and the PM machine is constant over a normal range of load. This is not the case with the series- and compound-field machines, where the field current and resulting field strength is a function of the machine load.

MOTOR REGULATION

The slope of the torque-speed characteristic of a motor is defined as its regulation, or $R_M = \Delta N/\Delta T = dn/dt$. This parameter has great significance for shunt-field and permanent-magnet motors, where the characteristic is a straight line. The regulation of a motor has additional significance in speed control applications.

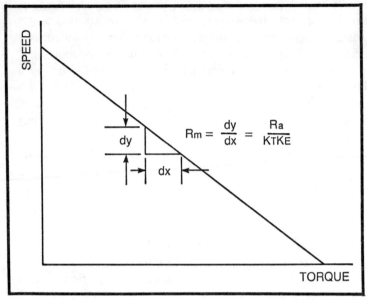

Fig. 5-4. Straight-line characteristics of the shunt field and/or PM motor.

The dimensions of motor regulation are units of speed per units of torque, i.e., [rpm/oz-in.] or [rpm/lb-ft]. This is an indicator of how a motor speed will fall off as load torque is applied.

A convenient expression can be derived directly from the torque-speed curve that will be applicable to shunt-field and PM motors. Refer to Fig. 5-4. Since the characteristic is linear, its equation will be that of a straight line with a negative slope, or

$$y = -mx + b \qquad (5.6)$$

The slope of the curve is seen to be

$$m = \frac{-N_0}{T_s}$$

and the constant, or Y intercept, is

$$b = N_0$$

If the variables, n and t are used in place of y and x respectively, Eq. 5.6 can be rewritten as

$$n = - \left(\frac{N_0}{T_s}\right) t + N_0 \qquad (5.7)$$

but

$$N_0 \approx \frac{V_1}{k_E} \tag{5.8}$$

and

$$T_S = \left(\frac{V_1}{R_1}\right) k_1$$

If the additional substitutions are made in Eq. 5.3, a new expression is obtained.

$$n = -\frac{R}{k_E \, k_1} \, t + \frac{V}{k}$$

then since, by definition

$$R_M = \frac{dn}{dt}$$

it follows that regulation can be expressed as

$$R_M = \frac{R_A}{k_E \, k_T} \tag{5.9}$$

$$n = N_0 - R_M t \tag{5.10}$$

If the regulation of a motor is known, the speed for any desired value of load can be easily calculated.

Example. A ¾ hp shunt-field motor has the following characteristics:

$$K_E = 0.050 \text{ [volts/rpm]},$$
$$k_T = 67.5 \text{ [oz in./amp]}$$
$$R_A = 0.65 \text{ [0hms]}$$

What will the speed be if a voltage of 90 volts is applied and the motor is loaded with 432 oz-in. of torque.

From Eq. 5.4

$$N_0 = \frac{90}{0.05} = 1800 \text{ [rmp]}$$

calculate R_M
from Eq. 5.5
$$R_M = \frac{0.65}{(0.05)(67.5)} = 0.192 \, \frac{\text{[rpm]}}{\text{oz-in.}}$$

Then motor speed is calculated form Eq. 5.10

$$n = 100 - (0.192)432$$
$$n = 1717 \text{ [MRPM]}$$

105

K_T \ K_E	$\dfrac{\text{oz-in.}}{\text{amp}}$	$\dfrac{\text{lb-ft}}{\text{amp}}$	$\dfrac{\text{lb-in.}}{\text{amp}}$
$\dfrac{\text{volts}}{\text{rpm}}$	1352	7.04	84
$\dfrac{\text{volts}}{\text{rad/sec}}$	141	7.3	8.8

Table 5-2. Torque-Voltage Conversion Constants

It should be noted that the regulation of a motor is temperature sensitive to the same degree that the factors which determine it are. Thus if significant temperature rise is associated with a particular operating condition, those thermal effects should be taken into consideration.

The temperature sensitivity of the torque-speed curve as shown in Fig. 5-3 can now be interpreted in terms of the machine characteristics just discussed. As the machine heats up from the power dissipated internally, the air-gap flux is reduced. The reduction in air-gap flux causes a proportionate reduction in the machine voltage constant. Equation 5.2 shows that for a given terminal voltage, the machine speed will increase if the voltage constant is reduced. The net result is a higher no-load speed as the motor heats up.

The same factors act to increase the regulation as temperature increases. Consider Eq. 5.5. The numerator on the right side (R_A) increases with temperature. At the same time both factors in the denominator will decrease in value with increasing temperature. This causes the slope of the torque-speed curve to change very rapidly as the motor warms up.

The full-load point is usually well to the left of the intersection point of the "cold" and "hot" curves. This means that motor speed is likely to increase significantly as the warm-up occurs.

INTERRELATIONSHIP OF TORQUE AND VOLTAGE CONSTANTS

If one looks closely at the equations for voltage constant (5.3) and torque constant (5.5), it is immediately evident that the same factors occur in both expressions. This commonality of factors makes it possible to derive an expression relating one characteristic to the other. This is done by dividing Eq. 5.5 by Eq. 5.3. If this is done, and the common terms cancelled, the expression obtained is,

$$k_T = k_2 k_E \qquad (5.11)$$

The value of the constant k_2 will depend upon the dimensional units of k_T and k_E as shown in Table 5-2.

ROTOR (ARMATURE) INERTIA

The inertia of a machines rotor is a property of the mass and its distribution in the rotor. Of particular interest when considering a machine's dynamic characteristics is the rotor polar moment of inertia. This property can be regarded as the ratio between an accelerating torque and the resultant acceleration (Fig. 5-5).

$$J = \frac{T}{\alpha} \qquad (5.12)$$

The moment of inertia of a machine has the utmost significance in determining what magnitude of speed changes will occur as a

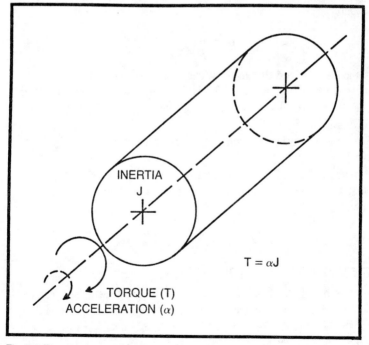

Fig. 5-5. The angular acceleration of a body is determined by its polar moment of inertia and the applied torque.

result of changes in shaft torque. It also is the determining factor in calculating the torque required to achieve a desired acceleration. Or conversely, it will determine the time required to achieve a desired speed if torque is limited by consideration of a maximum allowable current.

An earlier section described the great pains that are taken in servo motor designs to mimize inertia in order to achieve high responsiveness. The opposite measures may also be taken to maximize inertia when a constancy of speed is required. For example, if it is desired to hold a generator voltage exactly constant, the addition of flywheel inertia will overcome any torque perturbations which may be present.

Most manufacturers provide data describing the dynamic potential of their motors. This data is usually referred to as the WK^2 value and is provided in dimensions of [pound-ft^2]. It is necessary to divide this value by the gravitational constant g (32.2 ft/sec^2) in order to obtain the true inertia dimensions. The acceleration is then calculated directly from Eq. 5.13 in dimensions of radians per second per second.

$$\alpha = \frac{T}{WK^2/g} \qquad (5.13)$$

It should be noted that in most cases motors accelerate loads that also have significant inertia. When this is the case the motor WK^2 and load WK^2 are added to determine the total that will affect the acceleration.

Example. A motor is required to accelerate at a rate of 500 [rad/sec^2].
If the rotor has a WK^2 of 6[lb.ft^2] and the load has WK^2 of 8[lb.ft^2], what torque is needed to produce the required acceleration?
Solve Eq. 5.13 for T

$$T = 500 \times \frac{14}{32.2} = 217.4 \text{ lb.ft}$$

THE MECHANICAL TIME CONSTANT

Some of the machine parameters already discussed can be combined to provide a figure of merit as to the motors accelerating capability. This figure of merit is the motor mechanical time constant.

The physical significance of mechanical time constant can more readily be understood by referring to Fig. 5-6. As the motor acceler-

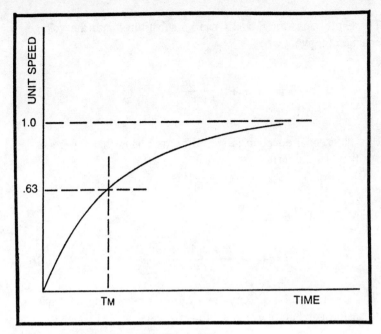

Fig. 5-6. Motor velocity profile after start-up.

ates under a constant torque, the speed increases along an exponential curve. The time at which the speed reaches 63% of its ultimate speed is defined as the mechanical time constant. If an expression for motor speed is derived it will be found to contain a time varying exponential term. The mechanical time constant will be the coefficient of time in the exponent.

The expression for time constant is derived as follows. First an expression is written for the voltage across the motor terminals. This equation will have two terms. One term represents the ohmic drop in the motor circuit while the second term represents the generated voltage.

Then

$$V_T = E_A + IR_A \tag{5.14}$$

but

$$E_A = k_E n$$

and

$$I = \frac{T}{k_1} = \frac{J_X}{k_T} = \frac{J\,dn}{k_1\,dt}$$

109

The expressions for E and I can be substituted into the voltage equation to give,

$$V_T = k_E n + \frac{R_A J}{k_T} \frac{dn}{dt} \tag{5.15}$$

The differential equation is solved by assuming that the general solution has the form

$$n(\pm) = A\epsilon^{st}$$

Then Eq. 5.14 becomes

$$V_T = k_E A \epsilon^{st} + \frac{R_A J}{k_T} \frac{d(A\alpha^{st})}{dt} \tag{5.16}$$

The constants of Eq. 5.16 are evaluated by writing its Laplace Transform and setting it equal to zero

$$0 = k_E A \epsilon^{st} + \frac{R_A J s A \epsilon^{st}}{k_T}$$

then

$$s = \frac{-k_T k_E}{J R_A} \tag{5.17}$$

where s is the complex frequency of the exponential function and is the reciprocal of the mechanical time constant.

$$T_M = 1/s = \frac{J R_A}{k_T k_E} = \frac{J R_A}{k_T k_E} \tag{5.18}$$

A new expression for motor speed can be obtained by solving Eq. 5.15 for n and substituting Eq. 5.17 as the value for s. When all the appropriate manipulations have been made, the final expression becomes

$$n = \frac{V_T}{k_E} - \exp\left(\frac{-k_T k_E t}{J R_2}\right)$$

but at $t = 0$, n is also zero, so

$$A = \frac{V_T}{k_E}$$

Then the final expression for motor speed can be written as

$$n = \frac{V_T}{k_E}\left[1 - \exp\left(\frac{-k_T k_E t}{J R_A}\right)\right] \tag{5.19}$$

Equation 5.19 is useful in providing an expression for calculating motor speed at any instant of time during the acceleration period.

Care must be taken to have all parameters in consistent units when using equations 5.17, 5.18, and 5.19. The voltage constant should be multiplied by the constant 0.1045 to change volts/rpm into volts per radian per second.

Example. Calculate the mechanical time constant for the motor with the following characteristics:

$$WK^2 = 0.071 \ [\text{lb}\cdot\text{ft}^2]$$
$$R_A = 1.5 \ [\text{ohms}]$$
$$K_E = 0.1 \ [\text{volts/rpm}]$$
$$K_T = 0.7 \ [\text{lb}\cdot\text{ft/amp}]$$

First find the polar moment of inertia

$$J = \frac{WK^2}{g} = \frac{0.071}{32.2} = 0.002 \ [\text{lb.ft.sec}^2]$$

Use Eq. 5.18 to calculate the time constant

$$T_M = \frac{(0.1045)(0.0022)(1.5)}{0.1 \times 0.7} = 0.0047 \text{ seconds}$$

Calculate motor speed that will exist 6 milliseconds after the application of 50 volts to the armature.

Use Eq. 5.19

$$n = \frac{50}{0.1} \quad 1 - \left[\left(-\frac{0.006}{0.0047}\right)\right]$$
$$n = 500 \ [1 - 0.271]$$
$$n = 364 \text{ rpm}$$

TEMPERATURE RISE

A dc machine converts energy from one form to another, electrical to mechanical and vice versa. The energy or power conversion process, however, is not achieved with a 100% efficiency. A certain percentage of the machine input is lost in various parts of the machine. Regardless of the nature of these losses, whether it be due to brush friction, hysteresis in the core iron, or joule loss in the conductor, all the lost power ultimately turns to heat. The heat generation in turn raises the temperature of the machine.

Because the insulating materials used in dc machines have definite temperature limitations, heat generation, cooling features, and temperature rise considerations are very important. Selecting a machine that is much larger than necessary to do a job is usually a safe way to go but certainly not the most economical. For that reason it is desirable to have an analytic procedure that provides a means of making approximate temperature calculations when duty cycle is time varying.

An approach that has been used over the years to predict temperature rise with a fair degree of success is the equivalent circuit method. This analytical method is based upon the similarity of the equations:

$$V = I \times R \text{ (Ohm's law)} \tag{5.20}$$

$$\Delta T = Q \times R_1 \text{ (Fourier's equation)} \tag{5.21}$$

Ohm's law is very familiar to motor designers and motor users and is immediately identified with the well developed analysis for electrical circuits. Fourier's equation relates the factors of heat transfer where ΔT is a temperature rise, Q is a quantity of thermal power to be dissipated, and R_1 is the thermal resistance of the transfer medium.

Figure 5-7 shows a typical situation encountered in heat transfer. The cross sectional view in (a) could represent the flow of steam through an insulated pipe. As the steam moves through the pipe some heat is transfered to the pipe from where it flows radially outward, conducted by the insulating material. The temperature profile shown in (b) shows the highest temperature at the pipe center. There is a slightly lower temperature at the pipe surface and then a steady fall to the outside temperature.

The similarity of Eqs. 5.20 and 5.21 permits the use of an analogue electrical circuit, which more readily displays the complex heat transfer in an electric machine. The "thermal terms" can be

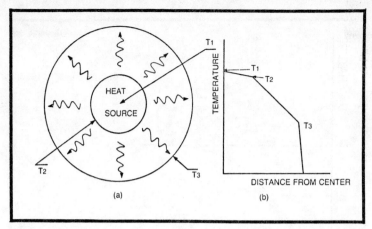

Fig. 5-7. Heat flow through a conducting medium, (a) the flow of heat is directed radially outward from the source to a cooler region and (b) the temperature profile for the heat flow shown in (a).

determined numerically and then plugged directly into the analogue circuit.

The circuit used to make temperature rise calculations can be very simple or quite complex, requiring the use of a digital computer for solution. If done carefully, however, even a simple calculation can yield useful results and will provide excellent insight into temperature factors.

In addition to the three factors shown in Eqs. 5.20 and 5.21 an additional energy storage element is needed. This is because when a machine is first excited, the temperature rises rather slowly along an exponential curve as shown in Fig. 5-8. As the temperature increases, heat energy is stored in the material of the motor, in particular the metal parts. The greater the amount of mass in a machine, the greater will be the thermal storage capacity.

The thermal capacity of a machine can be calculated if information is available as to its materials of construction. Thermal capacity is given by the expression,

$$h = M \times c \tag{5.22}$$

where h is the thermal capacity in dimensions of [watt•sec/°c]
M is the mass of the machine in [lbs].
c is the specific heat of the machine in [watt•sec/lb.°C]

Consider the dimensions of h in Eq. 5.22. It is seen that the following analogue equivalency exists,

$$[\text{watt•sec/°C}] \sim [\text{amp•sec/volt}] \tag{5.23}$$

113

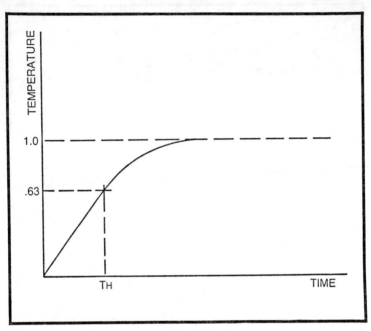

Fig. 5-8. Temperature rise vs. time in a dc machine for a constant level of power dissipation.

The right hand side of Eq. 5.23 represents dimensions of electrical capacitance. Then a capacitor can be used in the thermal circuit to represent the heat storage inherent in a machine. The analogue quantities are shown in Table 5-3.

AN EQUIVALENT ANALOGUE CIRCUIT FOR THERMAL ANALYSIS

In its simplest form an equivalent thermal circuit can be drawn as in Fig. 5-9. The circuit consists of a single current source, a single resistor, and a single capacitor.

Table 5-3. Thermal and Electrical Analogues

Thermal Property	Electrical Property
Temperature rise [°C]	Voltage [volts]
Heat rate [watts]	Current [amperes]
Thermal resistance [°C/watt]	Resistance [ohms]
Thermal capacity [watt • sec/°C]	Capacitance $\left[\dfrac{amp \cdot sec}{volt}\right]$

In this simple representation, all the machine losses are lumped together and regarded as originating from a single source. Similarly the thermal resistance and heat storage are assumed to be appropriately represented by a single element. The circuit shown in Fig. 5-9 does not provide a means for calculating the temperature at different parts of the motor. Yet even in its simplicity it does provide a means of approximating the hot spot temperature that will exist over short periods of time.

The circuit of Fig. 5-9 is particularly useful if some temperature test data is available on a motor. Most motor manufacturers will "heat-run" test a motor to make sure its operating temperature does not exceed its temperature rating. If this test data is available, it is possible to determine, a value for the resistor in Fig. 5-9.

$$R_T = \frac{\text{temperature rise}}{\text{machine losses}} \qquad (5.24)$$

The capacitor value can be calculated by multiplying the weights of the metals used in the motor by their respective specific heat coefficient. The individual heat storage elements are then summed to provide a single value.

$$C = M_I C_I + M_C C_C + M_A C_A + \ldots \qquad (5.25)$$

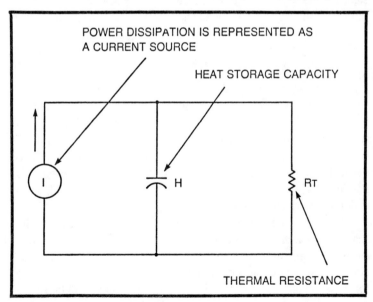

Fig. 5-9. Simple thermal circuit analogue representation of a dc motor.

The factors M_x represent the weights of iron, copper, aluminum, etc. used in the motor.

The factors C_x represent the specific heat for each of the specific metals. When specific information is lacking, an approximation of the mass weights can be made. Taking 85% of a machines weight as iron and 15% as copper gives acceptable results.

The current value to be used in the circuit is the total machine losses in watts.

The methods of circuit analysis can be used on the thermal circuit. The current response in the resistor of the circuit will be given by,

$$i = I \left[1 - \epsilon^{-t/R_TC} \right] \qquad (5.26)$$

The voltage drop across the resistor, which corresponds to the temperature rise, will be,

$$v = iR_1 = R_1I \left[1 - \epsilon^{-t/RC} \right] \qquad (5.27)$$

The similarity between Eq. 5.27 and Eq. 5.19 is easily recognized. This should not be too surprising since the characteristic curves of the two expressions had a similar exponential shape. The significance of this feature is that a dc machine does have a thermal time constant associated with it. The time constant is determined by the product of thermal resistance and heat storage capacity. Or

$$T_H = R_1 \times C \qquad (5.28)$$

The time constant is the period of time required to achieve 63% of an ultimate temperature rise at a constant rate of heat dissipation. Knowledge of a motor's thermal time constant is particularly useful in applying motors where overloads are required for short periods of time. If the required on-period is short compared to the machine thermal time constant, followed by a relatively long off period, it is possible that excessive temperatures may not be exceeded, even under an overload condition.

While the analogue circuit allows thermal problems to be analyzed by familiar circuit techniques, there are striking differences. For example the electrical time constant of a typical RC circuit is usually on the order of milliseconds. On the other hand, in the case of the thermal circuit, it is not unusual to obtain time constants of an hour or more.

Example. A motor weighing 26 pounds develops ¾ horsepower at 85% efficiency and has a stabilized temperature rise of 70° C. What is the thermal time constant? Calculate the temperature rise if operated for 20 minutes.

First determine the total loss in the motor:

$$P_L = \tfrac{3}{4}(746)\left(\frac{1}{0.85} - 1\right)$$

$$P_L = 98 \text{ watts}$$

Thermal resistance is calculated from Eq. 5.24.

$$R_T = \frac{70}{98} = 0.715° \text{ C/watt}$$

Heat storage capacity is calculated by approximating the weights of copper and steel.

weight of copper $= 0.15(26) = 3.9$ pounds
weight of steel $= 0.85(26) = 22.1$ pounds

then for the copper $C = 3.9$ [lb] × 174.5 [watt sec/lb°C] = 681 [watt · sec/°C] and for the steel $C = 22.1$ [lb] × 208.5 [watt · sec/ lb°C] = 4650 [watt · sec/°C] total heat storage is 5331 [watt · sec/°C] Then the thermal time constant is obtained by use of Eq. 5.28,

$T_H = 5331$ [watt · sec/°C] × 0.715 [°C/watt] = 3810 sec.

The temperature rise is calculated using Eq. 5.27 and a value of 20 minutes for the time.

$$V = (0.715)(98)\left[1 - \epsilon^{-(1200/3800)}\right]$$

$$V = 19°C$$

COMPLEX THERMAL CIRCUITS

The circuit of Fig. 5-9 is useful but provides a very limited amount of information. Typically it is desirable to know what temperatures might exist at different points within a machine. To obtain a more detailed picture of the internal temperature profile, a more complex circuit is required.

A more developed thermal circuit is shown in Fig. 5-10. The circuit of Fig. 5-10 requires detailed information of machine losses and internal features of construction. The various thermal resistances can be calculated from a theoretical approach, or determined empirically by instrumenting a machine with thermocouples and testing.

The circuit elements used in Fig. 5-10 are as follows:

Power Losses

- I_1 is the power loss at the brush-commutator interface.
- I_2 is the copper I^2R loss in the end-turn portion of the armature at the commutator end.
- I_3 is the copper I^2R loss of the armature conductor embedded in the slots.
- I_4 is the copper I^2R loss of the end-turn portion of the armature at the end opposite the commutator.
- I_5 is the armature core iron losses.
- I_6 is the field winding I^2R copper loss.
- I_7 is the field core iron loss.
- I_8 is the windage and friction loss.

Thermal Resistances

- R_1 is the thermal resistance representative of heat flow between the commutator and the internal air.
- R_2 is the thermal resistance between the armature end turns at the commutator end and the internal air.
- R_3 is the thermal resistance between the active portion of the armature conductor and the armature core.
- R_4 is the thermal resistance between the end-turn portion of the armature conductors (opposite the commutator end) and the internal air.
- R_5 is the thermal resistance between the commutator and the armature end turns.
- R_6 is the thermal resistance between the commutator end-turn portion of the armature and the active portion in the slot.
- R_7 is the thermal resistance between the active portion of the armature and the end turns opposite the commutator end.
- R_8 is the thermal resistance between the armature core and the internal air.
- R_9 and R_{12} are the thermal resistances between the end brackets and the ambient air.
- R_{10} and R_{11} are the thermal resistances between the end brackets and the internal air.
- R_{13} is the thermal resistance between the field windings and the internal air.
- R_{14} is the thermal resistance between the field core and the ambient air.

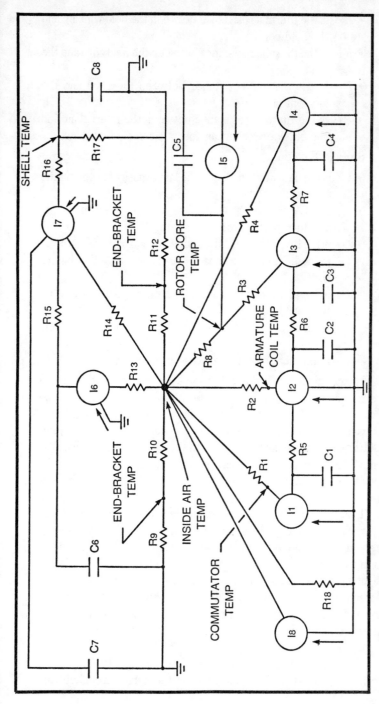

Fig. 5-10. Fully developed thermal circuit for more accurate temperature rise calculation.

119

- R_{15} is the thermal resistance between the field winding and the field core.
- R_{16} is the thermal resistance between the field core and the motor shell.
- R_{17} is the thermal resistance between the shell and the ambient air.
- R_{18} is the thermal resistance between the internal air and the outside ambient air for an air cooled machine.

Heat Storage Elements

- C_1 is the heat storage due to the large mass of copper in the commutator.
- C_2 and C_4 are the heat storage capacity inherent in the end-turn portions of the armature.
- C_3 is the heat storage of the active portion of the armature.
- C_5 is the heat storage capacity of the iron in the armature core.
- C_6 is the heat storage capacity of the copper used in the field winding.
- C_7 is the heat storage capacity of the iron used in the field core.
- C_8 is the heat storage capacity of the motor shell.

The circuit shown in Fig. 5-10 allows the calculation of temperature rise at specific points inside the motor. The solution of the circuit consists of finding the current distributions through the various available branches. After the current in each branch of the circuit is determined, it is easy to calculate the temperature at a particular point. The temperature rise above ambient consists of all the IR drops between the point of interest and ground (ambient temperature).

The circuit of Fig. 5-10 is a formidable one to work with without the availability of a digital computer with a suitable program. Fortunately, it is possible to greatly simplify the circuit without sacrificing too much in the way of accuracy. The value of Fig. 5-10 is in the insight it provides of the heat flow and heat storage that occurs inside an electrical machine. It is possible to see how the capacitors become fully charged as the temperature slowly builds up. In turn, as the heat transfer rate increases the temperature rise increases in direct proportion.

The circuit of Fig. 5-10 assumes a "mixing cup" condition of the internal air, i.e., the air temperature is the same throughout the machine. This assumption does not seriously undermine the value of the analysis since the hardware temperatures are of most interest.

The circuit can be used to represent either totally enclosed nonventilated (TENV) or open- and fan-cooled frames. If the machine is TENV, the branch containing R_{18} is merely opened. Of course, the values of all the thermal resistances between a machine part and the air is a function of the air flow over it. If the air is still, the heat transfer coefficients are at a maximum. While if there is moving air over a surface, the heat transfer coefficient decreases as air flow increases.

A SIMPLIFIED THERMAL CIRCUIT

Figure 5-11 shows a thermal circuit that is greatly reduced in complexity from Fig. 5-10, but still provides useful information. Its simplicity allows quick hand calculations without creating a dependency upon a computer.

In Fig. 5-11, the circuit elements represent the following quantities:

- I_1 is the total loss associated with the armature, including I^2R loss, iron loss, and commutator contact loss.
- I_2 is the total loss associated with the field including both I^2R loss and iron loss.
- C_1 is the total heat storage in the armature for both copper and iron materials.
- C_2 is the total heat storage in the field structure and the shell, including all copper and iron.
- R_1 is the thermal resistance between the armature and the internal air.
- R_2 is the thermal resistance between the field winding and the internal air.
- R_3 is the thermal resistance between the inside of the motor and the outside ambient.

Figure 5-11 shows a significant fact about a machine merely by a glance. It shows that both the armature and the field have their own unique thermal time constant associated with them. For the armature, the thermal time constant is given by R_1C_1 while for the field it is given by $(R_2 + R_3)C_2$. The relative values of these time constants weighs very heavily especially when the armature has a very low time constant with respect to the field. This situation does occur in low inertia servo motors. As a result, it is possible for an overload

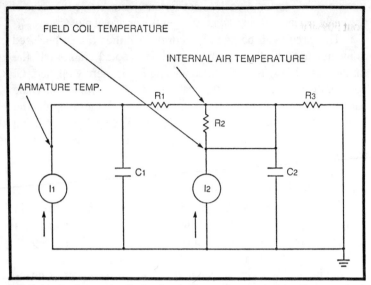

FIELD COIL TEMPERATURE

INTERNAL AIR TEMPERATURE

ARMATURE TEMP.

Fig. 5-11. Analogue circuit showing both armature and stator time constant effects.

condition of less than 30 seconds to destroy the armature while barely raising the temperature of the surrounding field by more than a few degrees.

The circuit of Fig. 5-11 can be easily calibrated with test data from an existing machine. This is done by measuring stabilized temperature rise at the points of interest in the machine under different conditions of machine loss. The temperature rises allow the determination of the thermal resistances by use of Eq. 5.24. Determination of the heat storage elements is based upon knowledge of the respective mass weights as previously indicated.

COOLING AND THE MACHINE ENCLOSURE

The Complex Thermal Circuits section made brief mention of the fact that thermal resistances are reduced when air is made to move over the heated surface. This fact has great significance in the choice of a machine frame, open or enclosed. The enclosed machine depends upon natural convection to move heat from the inner parts to the outside. If this inefficient means of heat transfer is assisted by the use of an internal fan there will be some improvement in the cooling. In addition, if holes are placed in the enclosure of the machine, allowing free interchange of air between the ambient and the interior, the results are dramatic.

In general the power rating of a machine can be increased by about 50% if it is put into an open, fan cooled enclosure. Thus, a ½ hp enclosed motor can usually be used at ¾ hp with open frame and fan cooling. If the application conditions are not severe enough to demand an enclosed motor, it is good economics to use one that is ventilated.

VOLTAGE AND TORQUE RIPPLE

In Chapter 2 equations were derived for calculating the voltage and torque of a dc machine. The equations so derived were expressed as dc value by integrating over a complete cycle and then averaging. In this form they give no indication of the cyclical nature of the generated voltages and torques. The voltage and also the torque that is associated with a particular coil is a function of its angular position. It varies in an approximate sinusoidal manner as the coil rotates through the magnetic field.

A single coil armature is shown in Fig. 5-12. As this single coil armature is rotated through a complete revolution at a constant speed, the generated voltage will be time varying. If the voltage of the coil were measured through brushes and slip rings, the voltage would appear as in Fig. 5-13(a). The zero voltage portion of the curve occurs for the period of time that the coil sides are rotating through the interpole space. During this period of time there is no change in the magnetic flux linking with the coil and consequently no generated voltage.

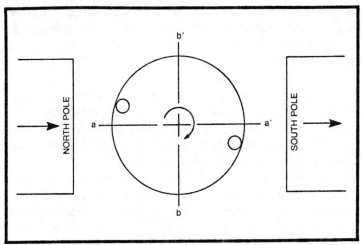

Fig. 5-12. A single coil armature machine.

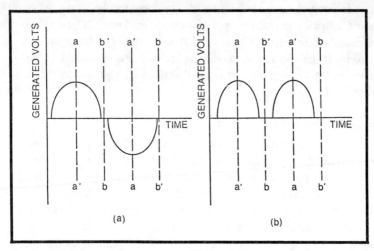

Fig. 5-13. Voltage waveforms generated in the single coil machine, (a) the coil voltage and (b) the terminal voltage.

The generated voltage is seen to be alternating in character with the polarity actually changing as the coil moves from one magnetic pole to another. Thus the coil voltage has a frequency which is given by the expression

$$f = \frac{Pn}{120} \text{ [hertz]} \tag{5.29}$$

Figure 5-13(b) shows the voltage measured at the brushes when the coil is connected to a commutator. As has already been discussed, the commutator serves to rectify the alternating coil voltages so as to produce a unidirectional dc voltage. However, the voltage, although dc, remains pulsating with a peak value and a zero for each pole of the machine.

A practical machine consists of many coils uniformly distributed about the periphery of the rotor. When these coils are appropriately connected, their instantaneous voltage add to one another. Because of the pulsating nature of the individual coil voltages, however, there remains a trace of time variation in the output.

The output voltage of a multicoil armature is shown in Fig. 5-14. It should be kept in mind that the terminal voltage is actually the summation of the instantaneous coil voltages at a particular instant of time. The output voltage of a generator can be characterized by a ripple specification. The ripple is expressed by the following expression,

$$\text{ripple voltage} = \frac{\text{maximum voltage} - \text{minimum voltage} \times 100\%}{\text{average voltage}}$$

$$(5.30)$$

Ripple characteristic is improved by increasing the number of voltage generating coils. This also requires a proportionate increase in the number of commutator bars and as might be guessed, acts to increase manufacturing costs. Nevertheless many applications require very low ripple characteristics and justify the added cost features to achieve them.

The output torque of a dc motor has a similarly pulsating origin in the individual coils. When many coils are integrated into a complete armature the results are as shown in Fig. 5-15. The ripple content in motor torque can also be expressed as a percentage of the average value by Eq. 5.31:

$$\text{ripple torque} = \frac{\text{maximum torque} - \text{minimum torque} \times 100\%}{\text{average torque}}$$

$$(5.31)$$

Fortunately, torque ripple in a motor is difficult to detect in most cases. This is due to the smoothing effect resulting from the inertias of the motor rotor and the load. As the motor slows at a minimum torque value, the inertia of the rotating system provides

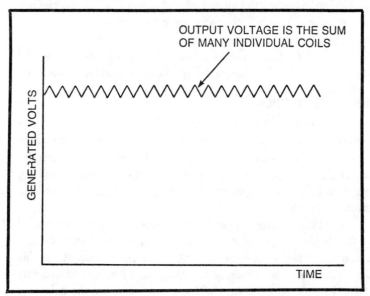

OUTPUT VOLTAGE IS THE SUM OF MANY INDIVIDUAL COILS

GENERATED VOLTS

TIME

Fig. 5-14. The output voltage waveform of a machine with an armature composed of many coils.

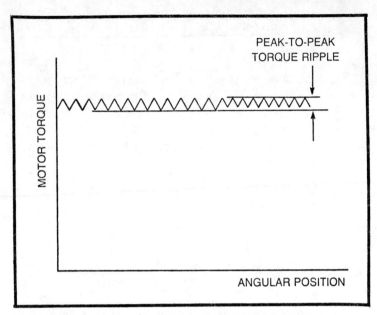

Fig. 5-15. Torque ripple characteristic of a machine with many coils.

power to the output shaft. As the motor torque increases to a maximum value, the increased torque causes the rotor to accelerate once more, replacing the stored inertial power. In this way the net output torque from a motor will mask the ripple inherent in the developed torque. But torque ripple does occur and in very sensitive servo speed control systems it may exceed the allowable speed variation. If such is the case, special design provision must be made.

ARMATURE REACTION

The various coil windings in a dc machine have as their purpose the creation of a magnetic field and may be regarded as electromagnets. In the same manner in which the field windings establish the main field or air-gap field, so does the armature winding establish a magnetic field. This magnetizing effect of the armature is called the armature reaction.

The field winding is usually concentrated in a single coil around each pole piece. The armature winding, on the other hand, is distributed in a series of coils around the entire rotor. Because of this distribution of coils, armature reaction is not confined to a clearly defined sector as is the case with the main field. On the contrary, armature reaction has the orientation as shown in Fig. 5-16.

126

The axis of armature reaction is established by the position of the brushes. Usually the brushes are placed at the quadrature axis (at 90° with the main axis). If such is the case, armature reaction is also in quadrature with the main field and it will have no components acting on the main field axis as shown in Fig. 5-16(b). However, Fig. 5-15(b) also discloses that at any angular displacement away from the main axis there is a component of armature reaction. The armature reaction will tend to either reinforce or oppose the effect of the main field.

Under normal load conditions, armature reaction is not a problem. Yet there are certain conditions under which it does constitute a definite machine limitation. For example, consider the condition where armature reaction is large with respect to the main field. If this condition exists, that portion of the air gap where the main field and the armature reaction oppose, will become demagnetized. Under such conditions the magnetic field in the air gap of the machine becomes very distorted. The consequences are likely to be severe sparking at the brushes and erratic machine performance.

AIR-GAP DEMAGNETIZATION IN A SHUNT/FIELD MOTOR

The shunt-field machine is a type in which the effects of armature reaction can give rise to a serious quirk in machine performance.

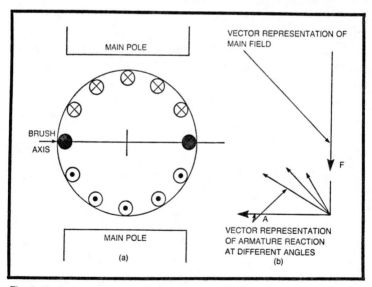

Fig. 5-16. Armature reaction in a dc machine.

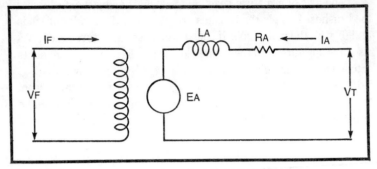

Fig. 5-17. Schematic of separately excited shunt-field motor.

The shunt field machine has a winding schematic as shown in Fig. 5-17. The field current is seen to have a constant value determined by the applied voltage and the winding resistance. The armature current, however, is a variable. Its value will depend upon the load torque applied to its shaft according to Eq. 2.14.

$$T = \frac{PZ\Phi I}{2\pi a} = K\Phi I_A \qquad (5.32)$$

Thus if the torque on a motor increases, the armature current will also increase. Along with the increase in armature current there will be an increased armature reaction. If the shunt field strength is insufficient for the value of armature current that opposes it, the resultant armature reaction will reduce the air-gap flux (the parameter Φ in Eq. 5.32). As can be seen from Eq. 5.32 the armature current will vary inversely with air-gap flux. Then as air-gap flux is reduced the armature current keeps increasing and producing a still stronger armature reaction to further reduce air-gap flux.

In order to fully appreciate the effect on motor performance, voltage Eq. 2.10 must also be considered.

$$E = \frac{PZ\Phi n}{60a}$$

or

$$n = \frac{KE}{\Phi} \qquad (5.33)$$

Rearranging the terms as in Eq. 5.33 shows that motor speed also varies inversely with the air-gap flux. Thus, as the air-gap flux is being reduced by armature reaction the motor continues to increase its speed.

The demagnetizing effect of armature reaction is a form of instability which can cause disastrous results to the machine. The torque-speed characteristic of an unstable machine is shown in Fig. 5-18. It can be seen that the curve has a positive slope, or negative damping. As torque increases, the speed also increases. The rotor of an unstable machine will race away to self-destruction. The speed continues to increase until the stresses created by the centrifugal forces exceed the mechanical strength of the rotor and it is torn apart.

An unstable condition exists with a shunt-field machine when the electromagnet represented by the armature becomes strong compared to the electromagnet of the main field. To prevent this condition from occuring, the main field winding must have a very high ampere-turn strength. Limiting the allowable motor overload will also serve to limit armature current and the resulting armature reaction.

Another design approach to avoid speed instability is the use of a stabilizing field winding. Such a field arrangement actually consti-

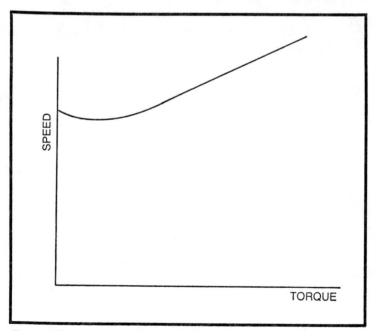

Fig. 5-18. Torque-speed characteristic of an unstable shunt motor.

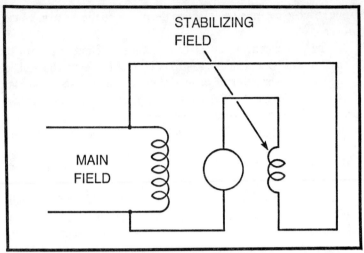

Fig. 5-19. Schematic of shunt-field motor with stabilizing field connection.

tutes a compound-field connection. A field winding of relatively few turns of large wire size is connected in series with the armature. Thus as armature current increases the same current flows through the series field winding and partially offsets the effects of armature reaction. The winding schematic of a motor with a stabilizing field is shown in Fig. 5-19. A stabilized motor will be capable of larger values of overload without danger of "running away."

Chapter 6
Methods of
Performance Analysis

Over the years a very comprehensive analysis has been developed which enables one to calculate and predict the performance of a machine at any condition of load. Techniques exist for making both steady-state and transient analysis. In this chapter a number of these analytical methods are explored.

THE SHUNT-FIELD MOTOR EQUIVALENT CIRCUIT

A very useful technique is to represent the shunt-field motor (or generator) by its equivalent circuit. This approach to analysis allows the well developed methods of linear circuit analysis to be used.

When developing the equivalent circuit to represent a machine, recognition is made of the fact that the armature winding is in fact a voltage generator. The generated voltage is given by the expression,

$$E_A = K_E n \tag{6.1}$$

The armature is also characterized by an ohmic resistance, (R_A) and if it is wound on an iron core it will also have a significant inductance.

The field winding is also characterized by ohmic resistance, (R_F) and an inductance (L_F).

The necessary circuit elements are then merely connected as in Fig. 6-1 to provide an equivalent circuit. The equivalent circuit as shown is elementary in that it does not account for all machine

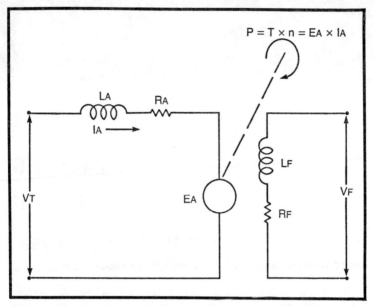

Fig. 6-1. The equivalent circuit representation of a shunt-field machine.

losses. However, it does permit easy calculation of the approximate torque-speed-current characteristics.

Then, if the rotational loss is assumed to be proportional to speed, it can be represented as

$$P_{ROT} = Dn \tag{6.2}$$

The developed electromagnetic power is given by

$$P_{EM} = E_A \times I_A \tag{6.3}$$

Net output power is equal to the developed power minus the rotational loss

$$P = P_{EM} - P_{ROT} = E_A \times I_A - Dn \tag{6.4}$$

Armature current is given by

$$I_A = \frac{T}{K_E K} + \frac{Dn}{V_1} \tag{6.5}$$

where K is a dimensional constant depending upon the units of torque used. (See Table 5-2).

Rotational speed is

$$n = \frac{V_T - I_A R_A}{K_E} \tag{6.6}$$

132

The total input power is,

$$P_{IN} = V_T I_A + V_F I_F \qquad (6.7)$$

Example. A 10 hp, 180 volt, 1750 rpm motor, has the following machine characteristics:

K_E = 0.099 volts/rpm
R_A = 0.159 ohm
D = 0.3 watts/rpm
R_F = 200 ohms

Calculate the current and speed when an overload condition exists which amounts to 150% of full load. Calculate the efficiency under over-load condition.

At 1750 rpm the full-load torque of a 10 hp motor is

$$T = \frac{10(746)\ 7.04}{1,750} = 30\ \text{lb.ft}$$

Then 150% of full load, torque is

$$T = 1.5(30) = 45$$

Use Eq. 6.5 to calculate armature current (Using K = 7.04) for consistency with the units of pound feet).

$$I = \frac{45}{7.04\ (0.099)} = \frac{0.3(1,750)}{180} = 67.5\ \text{amp}$$

$$n = \frac{180 - 67.5(0.159)}{0.099} = 1,712$$

The efficiency is calculated by first finding the output power and then dividing by the total input.

$$P = \frac{45 \times 1,711}{7.04} = 10,931\ \text{watts}$$

The input power is found from Eq. 6.7:

$$P_{IN} = 180(67.5) + \frac{(180)^2}{200} = 12,322$$

Then the efficiency is:

$$n = \frac{10,930}{12,322} \times 100\% = 88.8\%$$

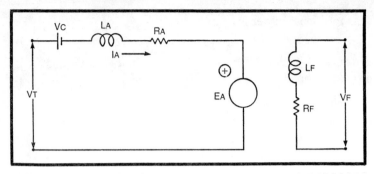

Fig. 6-2. The equivalent circuit representation including the brush contact drop, **V.**

IMPROVED EQUIVALENT CIRCUIT

The equivalent circuit just developed neglected to take into consideration a significant component of motor loss, viz, the brush contact loss. This loss, which is associated with the heat generated between the brush and commutator is characterized as a $V \times I$ power loss. For a reasonable range of current values around the full-load point, the contact drop is regarded as a constant. The contact loss, as a result, is proportional to the armature current.

If this additional feature is added to the equivalent circuit, the result is as shown in Fig. 6-2. Performance calculations based upon the circuit of this equivalent circuit will provide a closer correlation with test results.

When using the equivalent circuit with contact drop, it is necessary to modify the expression for motor speed as given by Eq. 6.6. If this is done, the new expression becomes,

$$ n = \frac{V_1 - V_C - I_A R_A}{K_E} \tag{6.8} $$

All the other analytical expressions previously developed are applicable as is. The effect of taking contact drop into consideration can be more fully appreciated by reconsidering the example calculation of the previous article.

Example. If the 10 hp motor previously analyzed has a brush contact drop of 4 volts, calculate the same parameters as done previously.

The current required for the overload condition remains the same at 67.5 amps.

The speed is calculated at

$$n = \frac{180 - 4 - 67.5(0.159)}{0.099} = 1670 \text{ rpm}$$

The output power becomes

$$P = \frac{45 \times 1,670}{7.04} = 10,690$$

The total input power remains the same at 12322 watts. The efficiency is,

$$n = \frac{10,690}{12,322} \times 100\% = 86.7\%$$

It can be seen that the effect of including brush contact drop produces a measurable difference in characteristics. An even greater significance is that of the impact of the additional losses on the motor temperature rise. So, while the difference in motor input/output characteristics may not seem particularly noteworthy, it should also be recognized that the motor losses have increased much more dramatically.

In the case where the brush contact drop was not considered, the total machine losses were, $12,322 - 10,930 = 1,390$ watts. In the second case where contact drop was considered, the total losses were found to be, $12,322 - 10,690 = 1,632$ watts. The total machine losses had thus increased by some 17%. This would be likely to produce a proportionate increase in machine temperature rise with possible disastrous effects.

It can be seen that the simple equivalent circuit of Fig. 6-1 provides reasonably accurate results for calculating current, torque, speed, and voltage. To provide an indication of the machine heating effects, though, the more complete circuit of Fig. 6-2 should be used.

The circuit elements shown in the equivalent circuits are lumped parameters. The circuit element R, when used in this type of analysis, must include all of the resistance of the armature circuit. It will, therefore, include any interpole and/or compensating winding resistances which may be present.

THE EQUIVALENT CIRCUIT OF THE PM MOTOR

The PM motor also lends itself to an equivalent circuit analysis. The expressions developed in the previous sections for the constant-field shunt motor are also valid for the constant field PM motor. The only difference between the analysis of the two motor types is that the PM motor does not have the field winding analogy or its accompanying power dissipation.

THE COMPLEX IMPEDANCE OF DC MACHINES

The equivalent circuits for both the shunt-field motor and the PM motor include the inductance of the winding. This parameter is not significant in most steady state modes of application. But it does take on added significance if the application is such that rapid starts and stops are required or if the dc power is provided by a chopper or some other type of switching device.

It should be kept in mind that motor action is a result of interaction between a current and a magnetic field. Therefore, any factor which slows the response of armature current will have an attenuating effect on the resulting motor action. For this reason the inductance of the armature is an important parameter and in some cases will effectively limit the dc current to a lesser value than that predicted by the winding resistance alone.

The current response of the armature circuit is determined by its electrical time constant, which is the ratio of inductance to resistance, L_A/R_A. The current response of an inductive circuit is shown in Fig. 6-3.

If the electrical time constant is significant when compared to the on time of an interrupted voltage, such as from phase controlled rectified ac or chopped dc, the dc value of armature current will be influenced by the inductance. If the electrical time constant is an order of magnitude less than the on time of the voltage the effect of the inductance can be neglected.

The complex impedance characteristics of an armature winding become very significant when it is applied as a high performance servo. A later section in this chapter will develop transfer functions for the dc motor. It will then be seen that the electrical time constant appears as a factor in the transfer function and influences transient response.

A later chapter will develop in greater detail the role inductance plays in limiting the high speed performance of motors when used with SCR phase controlled rectified ac.

THE SERIES AND COMPOUND FIELD MOTORS

The equivalent circuit analysis is only convenient for machines that have an essentially constant value of air-gap magnetic flux. This

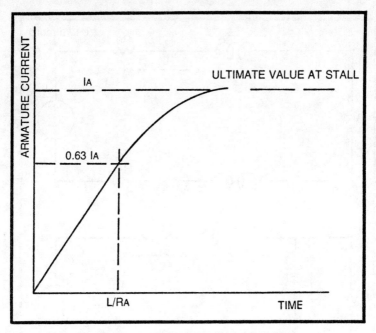

Fig. 6-3. Current response in the windings of the dc machine.

condition does exist with the shunt-field and PM machines but does not with the series- and compound-field machines. The latter two machine types have a field coil which carries armature current and as a result the field strength is a function of the current and motor load.

The actual value of the magnetic flux in a particular machine will also be greatly influenced by the degree of saturation of its magnetic circuit. This saturation effect in particular makes it difficult to predict machine performance without specific detailed information about the machine in question. For these reasons the equivalent circuit method is not as convenient for a motor with a series-field winding.

Nevertheless, an equivalent circuit can be synthesized if enough data is available. Circuits for series and compound motors are shown in Fig. 6-4. In general a saturation curve is required for calculating the performance of a motor with a series-connected field winding. Such a curve is shown in Fig. 6-5. It shows the air-gap flux that will exist for a given level of field excitation. If a saturation curve is available for a particular design, a torque-speed curve can be constructed by straight-forward calculations. A value of current is first assumed. The corresponding value of air-gap flux is then read from the saturation curve. With the level of air-gap flux known the

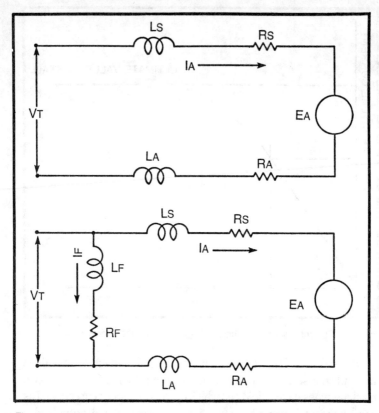

Fig. 6-4. Equivalent circuits for machines with series-connected fields, (a) straight series field and (b) compound field.

voltage and torque Eq. 2.10 and 2.14 are used to calculate the speed and torque corresponding to the assumed current. The process can be repeated for a number of points so that a smooth curve can be drawn. Of course the process of calculating the motor characteristic is greatly simplified if a computer and appropriate program are available for use.

If an actual machine is available it is possible to acquire by two simple test procedures data that will then allow calculation of performance over a wide range of application. The method works as follows.

The motor is driven at no-load by slowly increasing the voltage. At regular intervals the speed, current, and voltage is measured. The product of current and voltage provides the rotational loss of the machine. The rotational loss characteristic is then plotted as in Fig. 6-6.

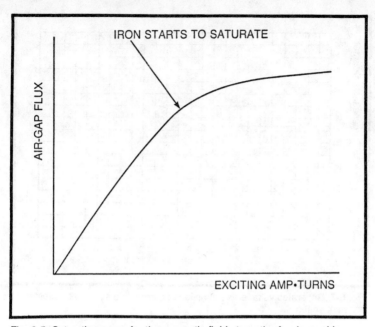

Fig. 6-5. Saturation curve for the magnetic field strength of a dc machine.

The machine is then tested in a generator mode by driving it with a constant speed motor. Care must be taken to reconnect any commutation compensating windings when this test is made. The machine is connected to a variable resistance load which is then

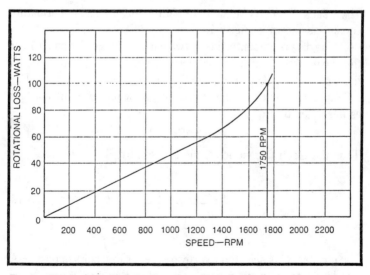

Fig. 6-6. The rotational losses of a series-field machine.

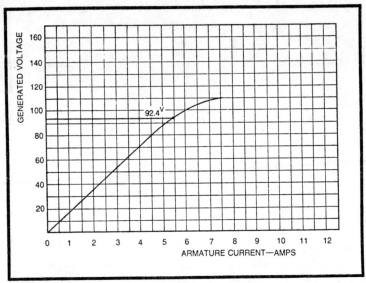

Fig. 6-7. Generated voltage as a function of current in a series motor tested at constant speed.

adjusted to provide different values of armature current. At each level of armature current the test machine terminal voltage is measured. The generated voltage is then determined by the relationship,

$$E_A = V_T + I_A R_A \qquad (6.9)$$

The values of E_A so obtained are plotted as a function of armature current as shown in Fig. 6-7.

The data represented in the two curves of Figs. 6-6 and 6-7 can then be used to calculate a complete torque speed characteristic for the test machine. The product of the coordinates of the curve in Fig. 6-7 represent the developed electromagnetic power of the machine as described by Eq. 6.3:

$$P_{EM} = E_A \times I_A$$

Then the electromagnetic torque is given by,

$$T = \frac{CP_{EM}}{n} = \frac{C\,(E_A \times I_A)}{n} \qquad (6.10)$$

The constant C is selected to provide the desired units of torque. A plot of electromagnetic power is also shown on Fig. 6-6.

The curve of Fig. 6-7 is used to find the electromagnetic power at a particular current level. Equation 6.10 is then used to calculate the equivalent torque for that value of current. The speed at any level of applied voltage and torque load can then be calculated using Eq. 6.9.

The process can be repeated for a number of current values to yield a smooth torque-speed characteristic.

Example. A series motor has the characteristics shown in Figs. 6-6 and 6-7. The generated voltage curve was obtained at a test speed of 1150 rpm. The total resistance in the armature circuit is:

R_A = 4.6 ohms at operating temperature

Calculate the speed and torque that will exist when the motor is excited with 180 volts and loaded to a current of 5.5 amperes.

Use Fig. 6-7, with a current of 5.5 amps, electromagnetic power of 508 watts, and an E of 92.4 volts at a test speed of 1150 rpm.

Then, with 180 terminal volts the generated voltage is from Eq. 6.9:

$$E_A = 180 - 5.5(4.6) = 154.7 \text{ volts}$$

The electromagnetic power is,

$$5.5 \times 154.7 = 851 \text{ watts}$$

The speed will be proportional to the generated voltage. So for the current condition of 5.5 amps if a speed of 1150 rpm produces 92.4 volts, a voltage of 154.7 means

$$n = \left(\frac{154.7}{92.4} \right) 1,150 = 1,925 \text{ rpm}$$

At this speed the rotational loss from Fig. 6-5 is 100 watts. Then the net output power is:

$$P = 851 - 100 = 751 \text{ watts}$$

The net available torque is:

$$T = \frac{751 \times 7.04}{1,925} = 2.75 \text{ pound·feet}$$

RMS TORQUE AND CURRENT

A large number of motor applications involve loads (or torques) that are time varying in nature. An example of such a varying load might be a compressor where the load is a function of a piston displacement. Or, an example of varying torque requirement is the servo motor that is required to accelerate an inertial load to a given speed and then decelerate at a controlled rate. Figure 6-8 shows examples of such time varying motor requirements.

The necessity often arises to calculate the heating effect produced by a time varying load. There might be a question of whether a current higher than the full-load value can be sustained for short but repetitive periods of time. In order to estimate the heating effect it is necessary to be able to calculate the joule heat in the windings.

The joule heating may be calculated by first calculating an rms torque required for the specific duty cycle. The approach is to look at the torque duty cycle and select a function that approximates it. The torque profile shown in Fig. 6-8 could be approximated by a sine function. If a sine is found to be appropriate, the rms torque would then have a value given by the peak torque divided by the square root of 2.

The torque profile shown in Fig. 6-8(a) is a series of constant torques of different values and for varying periods. The rms torque of such an application requirement is calculated by the following formula.

$$T_{RMS} = \sqrt{\frac{(T_1)^2\, t_1 + (T_2)^2\, t_2 + (T_3)^2\, t_3}{t}} \qquad (6.11)$$

Example. A dc servo motor is required to drive the tape reel for a magnetic tape drive. It is required to accelerate the reel at a constant torque of 500 oz in. for 0.5 second, run at constant speed with 50 oz in. for a period of 1 second, and then decelerate with 450 oz in. of torque for a period of 0.5 second.

What is the rms torque requirement of the motor?

From the information provided the parameters of Eq. 6.11 are:

$$T_1 = 500,\ T_2 = 50,\ T_3 = 450$$
$$t_1 = 0.5,\ t_2 = 1,\ t_3 = 0.5,\ t = 2$$

The using Eq. 6.11:

$$T_{RMS} = \sqrt{\frac{(500)^2 \times 0.5 + (50)^2 + (450)^2 \times 0.5}{2}}$$

$$T_{RMS} = 477 \text{ ounce inches}$$

The value of rms torque required for an application is often substantially higher than the average value. Although the average torque determines the average current, it is the rms value that influences the heating.

142

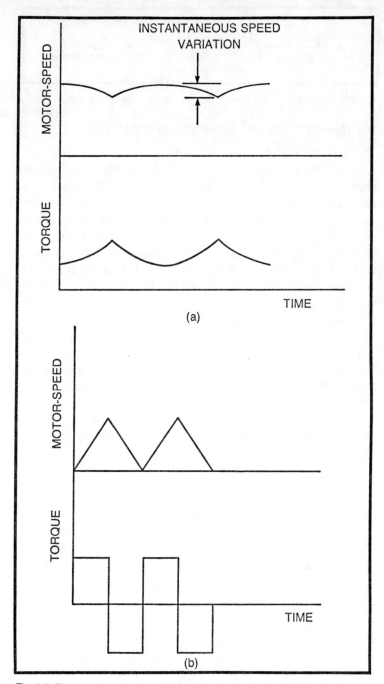

Fig. 6-8. Time varying motor torque requirements, (a) motor driving a compressor load and (b) servomotor driving a tape reel in a magnetic tape drive.

143

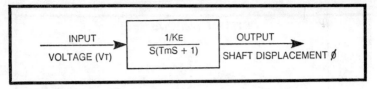

Fig. 6-9. Block diagram transfer function representation of dc motor.

THE TRANSFER FUNCTION OF SHUNT AND PM MOTORS

The transfer function representation of a dc shunt and permanent-magnet motor is very useful in the analysis of transient characteristics. It is also a very important tool available for analyzing instability in a motor. Or, conversely, the transfer function analysis allows the application of stability criterion to the whole design synthesis process. By means of this type of analysis, the engineer can interface the motor with other system components which are also represented in a transfer function form. In this way, the analysis of an entire system can be approached in a logical manner.

A transfer function is the ratio of an output to an input with the expressions written as LaPlace transforms. An elemental representation of a motor by means of a transfer function is shown in Fig. 6-9. This type of representation is a block diagram. The transfer functions of connected blocks will multiply. In this way the response (or output) of a complex system to a signal (or input) can be analyzed for a wide variety of parameters.

An expression for the transfer function is obtained as follows. The equation for the torque-speed characteristic is

$$n = -R_M t + N_0 \qquad (6.12)$$

If the shaft displacement is θ, then motor speed can also be written as

$$n = \frac{d\theta}{dt}$$

The torque can also be written in terms of machine parameters and output characteristics. If this is done,

$$t = J \frac{d^2\theta}{dt}$$

and

$$N_0 = \frac{V_T}{k_E}$$

144

If the appropriate substitutions are made, Eq. 6.12 can be rewritten as

$$\frac{d\theta}{dt} = -R \frac{Jd^2\theta}{dt} + \frac{V_1}{k_E} \qquad (6.13)$$

Equation 6.13 is rearranged and transformed to give

$$s\theta + R_M Js^2\theta = \frac{V_T}{k_E} \qquad (6.14)$$

The desired ratio is output (Ω) divided by the input (V). If Eq. 6.14 is solved for the ratio of output to input

$$\frac{\theta}{V_1} = \frac{1/k_E}{s + R_M Js} \qquad (6.15)$$

but

$$R_M J = \frac{R_A J}{k_E k_1} = T_M$$

(machines mechanical time constant, refer to Eq. 5.18)

So, Eq. 6.15 can then be rewritten as,

$$\frac{\theta}{V_1} = \frac{1/k}{s(T_M s + 1)} \qquad (6.16)$$

Equation 6.16 is the transfer function of a motor with the electrical impedance characteristics of the machine neglected. Usually this simplification is justified with the mechanical time constant an order of magnitude greater than the electrical time constant in an iron core machine. However, with certain types of servo motors the rotor inertia is minimized to such an extent that the two time constants associated with it (electrical and mechanical) approach a closer equity. In such a case it is necessary to include in the transfer function the electrical time constant term.

FREQUENCY RESPONSE OF THE MOTOR

If the transfer function of a motor is available it is an easy matter to analyze its response to signals that are of a cyclic or variable frequency nature. The frequency response is obtained by setting the complex variable s equal to $j\omega$.

Then

$$s = j\omega \qquad (6.17)$$

where j is the unit vector, $1\sqrt{90°}$, and $\omega = 2\pi F$

The identity of Eq. 6.17 is then substituted into Eq. 6.16 to obtain,

$$\frac{\theta}{V_T} = \frac{1/k_E}{j\omega\,(j\omega T_M + 1)}$$

This can be simplified to

$$\frac{\theta}{V_T} = \frac{1/k_E}{(-39.48\,f^2\,T_M + j\,6.28\,f\,)} \qquad (6.18)$$

In analyzing a complex system with a number of components represented by transfer functions it is convenient to use decibel notation. In this form the transfer functions may be added directly to each other to determine the system characteristics.

Equation 6.18 is written in decibel form by taking the logrithm of the right hand side and multiply it be 20. Then if Eq. 6.18 is written in decibel form,

$$\frac{\theta}{V_1} = 20\log\frac{1/k_E}{-39.48\,f^2\,T_M + j\,6.28\,f} \qquad (6.19)$$

The response of a particular machine can be seen graphically by inserting into Eq. 6.19 different values of f and making the prescribed calculations. If this process is followed for enough data points, a curve such as that shown in Fig. 6-10 can be drawn. The curve shows the output/input ratio for different values of frequency.

Two asymptotes may be drawn to the curve as in Fig. 6-10. If this is done it is seen that their point of intersection is at a point which is at an angular velocity equal to the reciprocal of the mechanical time constant. When viewed from this perspective of gain and bandwidth, it can be more fully appreciated that an electric motor also has a definite corner frequency associated with it which limits its range of possible applications. It is also possible to discern how the motor parameters by their effect on the mechanical time constant will affect the corner frequency.

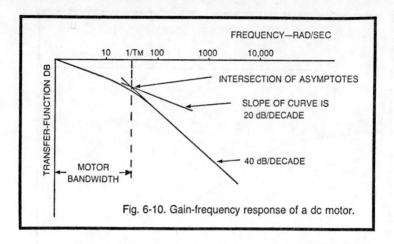

Fig. 6-10. Gain-frequency response of a dc motor.

The denominator of the transfer function shown in Eq. 6.16 is a complex expression whose terms do not add arithmetically. The effect of this is to cause a phase lag between the input signal and output response. The phase angle is given by

$$B = \tan^{-1} (T_M S) \qquad (6.20)$$

if

$$S = J\omega$$

it can be seen that the tangent of the angle and consequently the angle itself will increase with the frequency. It is also seen that when,

$$\omega = \frac{1}{T_M}$$

that

$$\tan B = -1$$

and

$$B = 135°$$

Thus, the phase angle at the corner frequency is 135° as would be expected.

The expression of Eq. 6.20 predicts a maximum phase angle of 180°. This would be the case if mechanical factors are the only ones considered. In fact, at higher frequencies the electrical impedance factors of the machine become increasingly important. The inductance of the armature will act to further attentuate motor response and increase the phase lag.

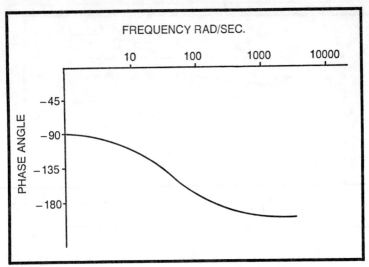

Fig. 6-11. Phase angle frequency characteristic of a dc motor.

Figure 6-11 shows a plot of phase lag versus frequency.

TRANSFER FUNCTION WITH ELECTRICAL TIME CONSTANT

A transfer function expression that takes into consideration the effects of armature induction can be derived by considering the circuit as shown in Fig. 6-1.

The voltage equation can be written as

$$V_T = R_A i + L \left(\frac{di}{dt} \right) + E_A \qquad (6.21)$$

but

$$i = \frac{T}{K_1} = \frac{J}{K_1} \left(\frac{d^2\theta}{dt^2} \right) + \frac{D}{K_1} \left(\frac{d\theta}{dt} \right)$$

and

$$\frac{di}{dt} = \frac{J}{K_1} \left(\frac{d^3\theta}{dt^3} \right) + \frac{D}{K} \frac{d^2\theta}{dt^2}$$

while

$$E_A = K_E n = K_E \quad \frac{d\theta}{dt}$$

If the appropriate substitutions are made in Eq. 6.21 and the equation transformed, it may be rewritten as,

$$V = \frac{R}{K_T}$$

$$(Js^2\theta + Ds\theta) = \frac{L}{K_T} (Js^3\theta + Ds^2\theta) + K_E S\theta.$$

The above equation is solved for the ratio θ/V_T to give (after simplifying identities are substituted),

$$\frac{\theta}{V_T} = \frac{1/k_E}{s[T_M T_E s^2 + T_M + T_E) s + 1]} \qquad (6.22)$$

or, in alternate form,

$$\frac{\theta}{V_T} = \frac{1/k_E}{s[(T_M s + 1) (T_E s + 1)]} \qquad (6.22a)$$

Equation 6.22 shows a second order term in the denominator indicating the significance of both mechanical and electrical delays to the response.

Chapter 7
Permanent-Magnet Machines

In the past two decades the PM machine has slowly evolved into an item of significant commercial importance in sizes ranging up to the small integral horsepower sizes. The trend in dc motor usage is definitely to the permanent magnet concept. It is quite likely that in another 10 years all but the most demanding applications for small motors will have gone to the PM machine.

There are a number of good reasons for the widespread proliferation of PM motors in the small sizes. The main reasons are merely listed here and will be discussed in greater detail later in the chapter. The reasons for the PM's popularity include:

- It requires a less complex and costly control.
- It offers definite performance advantages.
- It provides a given horsepower rating in a smaller package with less weight.
- It improves reliability by eliminating the field winding as a potential failure mode.
- It provides the user with the function he needs at lower cost.
- It has a "field on" condition with magnetic detent at all times, even in case of a power failure.

The modern PM motor would not have been possible without significant development of magnetic material characteristics.

PROPERTIES OF PERMANENT MAGNETS

A discussion of PM materials must be preceded by a description of the properties of a permanent magnet. Unfortunately, the field of magnetics has not progressed as far as some other areas of engineering in the use of one system of units. As a consequence data sheets describing magnet characteristics are more often seen using English or cgs (centimeter•gram•second) units rather than the preferred mksa (meter•kilogram•second•ampere) units. The result is an apparent confusion of terminology and the frequent need to convert from one system to another when doing analytical work.

The basic magnetic properties can be described by use of the following parameters.

Magnetic Flux

This describes the amount of magnetism, or lines of force, present. Throughout this text, the symbol Φ has been used to indicate magnetic flux. It is measured in units of:

- webers in the mksa system
- lines in the cgs system
- maxwells in the English system

Magnetomotive Force

This is a measure of the excitation that is acting to establish the magnetic flux. Since Ampere established at an early date that every current-carrying conductor had a magnetic field associated with it, magnetomotive force (mmf) is measured in terms of the product of a current and the number of turns of a coil. The units in the three systems are:

- ampere•turns in the mksa system
- gilberts in the cgs system
- ampere • turns in the English system

Magnetic Flux Density (or induction)

This is the amount of magnetism, or force lines, in a given area. In this text the symbol B is used to represent magnet flux density. The units of measurement are:

- webers/square meter in the mksa system
- gauss in the cgs system
- maxwells/square inch in the English system

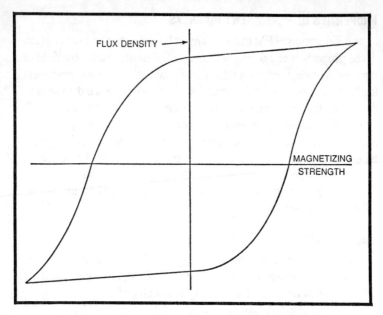

Fig. 7-1. The magnetic hysteresis of a permanent-magnet material.

Magnetizing Force

This is a measure of the intensity of the magnetomotive force acting at a point in a magnetic circuit. The symbol H is used to represent magnetizing force and the units of measure are:

- ampere·turns/meter in the mksa system
- oersteds in the cgs system
- ampere·turns/inch in the English system

The parameters defined above take on a special meaning when applied to permanent-magnet materials. Figure 7-1 shows the hysteresis curve of a PM material. It shows the relationship of magnet flux density to the magnetizing force in the four quadrants of the graph. The part of the hysteresis characteristic that is of special interest to this discussion is the curve which lies in the second quadrant. This portion of the hysteresis curve is redrawn in Fig. 7-2 with the points of interest labeled. This part of the hysteresis characteristic is known as the demagnetization curve of the permanent magnet.

The point B_R is called the residual flux density of the magnet. This is the value of flux density that will exist in a magnet that if fully charged, with no external demagnetizing force present.

The point H_C is called the coercive strength of the magnet. Its value gives the ampere-turn per inch of demagnetizing force that must be applied to the magnet to reduce the flux density to zero.

The portion of the curve between B_R and H_C shows the value of magnet flux density that will exist for a given value of demagnetizing force. Thus, if an external demagnetizing force, H_D, is applied to the magnet, the magnet flux density will decrease from B_R to a new value at B_D. The curve of Fig. 7-2 is called the normal demagnetization curve.

The value of B_R will only exist while the magnet is in a closed magnetic circuit as shown in Fig. 7-3. Any type of air gap in the magnetic circuit appears to the magnet as a demagnetizing force. The greater the air gap, the greater is the demagnetizing effect. Then since any practical PM motor must have an air gap for mechanical clearance, it can be seen that the magnet will be at a flux density less than the B_R associated with that material. The exact value of magnet flux density will be determined by the design of the air-gap.

THE INTRINSIC CURVE

The normal demagnetization curve shown in Fig. 7-2 can be thought of as consisting of two components. One is due to the

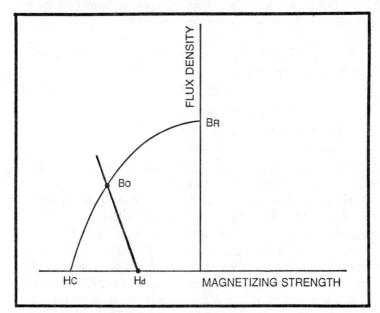

Fig. 7-2. The demagnetization curve of a PM material.

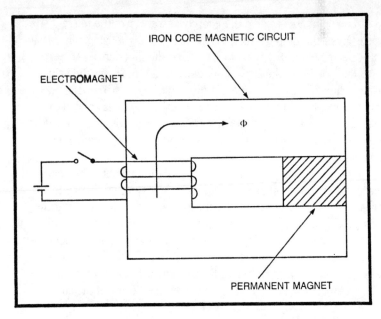

Fig. 7-3. A fully charged magnet in a closed magnetic circuit.

presence of the magnetic material itself. The second component is the flux density which would exist in free space if the magnet were removed. The flux density due to the magnet material itself is called the intrinsic induction of the magnet.

The normal curve is related to the intrinsic curve by the following mathematical relationship.

$$B = H + B_i \qquad (7.1)$$

where B is the normal characteristic,
$\quad\quad H$ is the magnetizing (or demagnetizing) force,
$\quad\quad B_i$ is the intrinsic characteristic.

(Equation 7.1 is written in cgs units because they happen to be the most convenient form for this expression.)

The physical concept of intrinsic induction can be a little nebulous. Still, it can be helped by the following explanation. Consider a tiny probe inserted into a cavity inside a fully charged magnet with an external field of magnetizing force H, as shown in Fig. 7-4. The probe will see an apparent magnetic charge on the surfaces of the cavity. It will also sense the external applied field, H. The total magnetic induction measured by the probe is then given by Eq. 7.1: $B = H + B_i$ Now if the external field is reversed but with reduced

154

strength, the condition is as shown in Fig. 7-5. The probe sees the same internal cavity surface charge, but now it sees a negative value of H external to the magnet.

This second condition as described in Fig. 7-5 is the one that exists in magnets when applied to dc motors. In this "second quadrant" condition with negative external H applied, it can be seen by rearranging the terms of Eq. 7.1 that $B_I = B + H$. Then if the intrinsic curve of the material described in Fig. 7-2 is constructed, it might appear as in Fig. 7-6. It is seen that the normal and intrinsic curves are common at point B_R. However, the point H_{CI}, which is called the intrinsic coercivity of the material, may be far greater than H_C. The intrinsic coercivity of a material has great significance. It is an indication of the material's ability to resist demagnetizing forces.

As will be seen, in addition to economic advantage, a principal feature of the ceramic family of magnets is the very high intrinsic coercivity associated with them.

Another point of interest is that the area under the normal curve has the units of energy per unit volume. A particular set of the curve's co-ordinates represents the energy storage, per unit volume, of the magnet. This can be seen by multiplying the coordinates of the two axes. If this is done,

$$b \times h = \text{lines/sq.inch} \times \text{ampere·turn/in.}$$

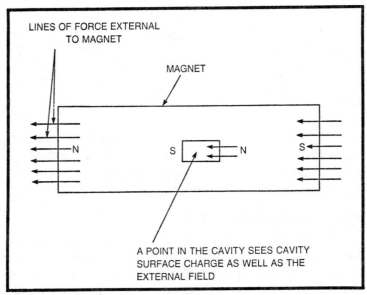

Fig. 7-4. The internal field of a small cavity in a magnet.

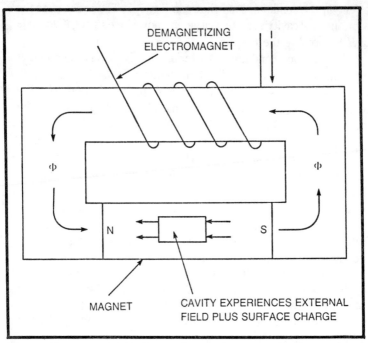

Fig. 7-5. The internal field of a small cavity in a magnet with an external demagnetizing force.

but

$$1 \text{ line} = \text{joule/ampere} \times 10^{-8}$$

so that

$$b \times h = \text{joule/cubic inch} \times 10^{-8}$$

The significance of the area under the curve gives rise to another figure of merit often applied to magnet materials. The product of the coordinates at which a maximum value occurs is called the energy product of the material. The continual evolution of new magnet materials has been characterized by ever increasing energy products. The significance of this figure of merit reflects on the amount of work a unit volume of the material can do.

MAGNET MATERIAL DEVELOPMENT

The development of permanent-magnet materials has proceeded along two main technological paths since the earliest days of electrical engineering. Today there are only two commercially important groups of permanent magnet materials. These are the alnicos and the ceramics. Steel magnets have ceased to be of commercial importance per se. Although chrome steel and cobalt steel materials are still used in a different mode in the rotors of hysteresis motors.

The cast alloys or alnico grades date from the early 1930s. The first alnico grades have been followed by subsequent improved grades. Each new development has shown an increase in the H_c or the B_R of the material. Figure 7-7 shows the demagnetization curves for a number of alnico materials.

The alnico grades have a number of very good characteristics from a design standpoint. They are only slightly affected by temperature. The B value of an alnico magnet will decrease by 0.01% to 0.02% per degree centigrade increase of temperature. This low level of temperature sensitivity means that PM motors using alnico magnets experience only a slight change in characteristics over a substantial range of temperature. (For example, if the magnet temperature changes by as much as 100°C, the change in magnet flux will only be on the order of 1.5%.

Of even greater motor-performance significance is the high level of magnetic induction obtainable from the alnicos. Alnico 5-7 has a B value of 13,500 gauss. This high level of induction means

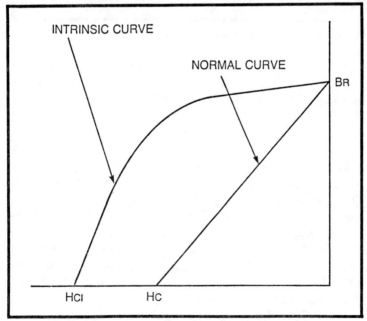

Fig. 7-6. The normal and intrinsic characteristics of a PM material.

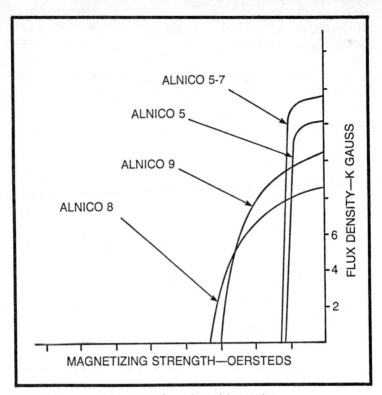

Fig. 7-7. Demagnetizing curves for various alnico grades.

many lines of force per magnet pole and a proportionately high level of motor torque development. For this reason the highest performance type of servo motor is built with alnico magnets.

The alnico magnets also have a number of serious deficiencies which are sufficient to have prevented it from taking over the dc motor field. Perhaps the largest deterrent to a mass use of alnico magnets is the high cost. As might be expected the cost increases with improvements in characteristics. Because of the material cost, alnico magnets are used only in those applications where the motor cost is but a small part of a piece of equipment whose function is dependent upon an ultra high performing motor.

The alnico magnets are costly because of the elaborate and carefully controlled processing that is required. Important constituent materials such as nickel, cobalt, and copper are of strategic importance and often in short supply.

A second severe limitation of the alnico class is its low coercivity, which translates into a susceptibility of demagnetization while in

158

application. The intrinsic curve is very rarely shown for an alnico material because it is only slightly different than the normal curve. This susceptibility to demagnetization requires extra care in the application of alnico PM motors. It is necessary to severely limit the current out of the controller to a safe value that will not cause demagnetization.

DEVELOPMENT OF CERAMIC MAGNETS

The development of the remarkable characteristics of the ceramic magnets has resulted from work done to develop powdered cores for choke coils. The purpose of these early core materials was to minimize the core losses that were associated with solid iron cores used in ac circuits. Oliver Heaviside experimented with cores made of iron particles as early as 1886.

As the technology of powdered core materials developed, equipment was built capable of grinding particles to very small and

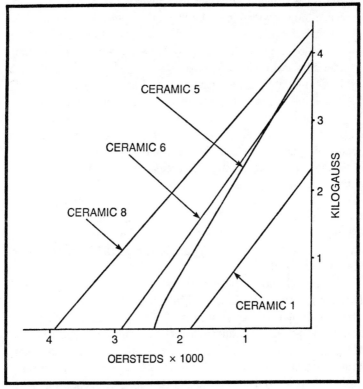

Fig. 7-8. Demagnetizing curves for various ceramic magnet grades.

uniform size. It was discovered that iron, which is not a permanent-magnet material in a solid macroscopic state, experienced an increase in its coercivity by several orders of magnitude when ground to a particle size of several microns' diameter. The discovery of these magnetic properties of iron particles led to the development of the ceramic or ferrite magnets.

Barium ferrite materials were investigated as early as the 1920s but did not receive the attention needed to fully develop their properties until the early 1950s. At that time a booming television industry created a great demand for a low cost focusing magnet. The research effort put into the ferrite powder materials resulted in the introduction of the first barium-ferrite permanent-magnet material about 1955. These magnet materials showed low B compared with the available alnico magnets. However, the coercivity (H) of the new material was far greater than anything else previously known. This produced a material that was capable of operating with very large air gaps and/or was highly resistant to demagnetization.

Since the introduction of the first grade of barium-ferrite, development has continued. As was the case with the alnicos, each newly developed grade has increased the energy product of the material, albeit with increased cost.

Barium-ferrite magnets have the mechanical characteristics of typical ceramics. They are good thermal and electrical insulators. They are lightweight, very hard, brittle, very weak in tension or compression. They will not corrode and cannot be drilled, cut, or ground by ordinary machine shop equipment.

Although the manufacturing process seems complicated, the magnets are made by molding and sintering, a process similar to that used to manufacture other ceramics. More recent oriented grades are pressed into form when in a strong magnetic field. The relative ease of manufacturing process when coupled with the fact that the raw material is cheap and readily available results in a magnet well suited for motor use.

Figure 7-9 shows the normal curves of a ceramic 7 magnet plotted on the same set of axes with an alnico 5-7 material. The great difference in their characteristics is readily apparent. The fact that the ceramic magnet has a linear characteristic makes it ideally suited for dynamic applications. The magnet might experience a slight temporary loss of magnetism if subjected to a very strong external field. If the external field is removed, though, the magnet will make full recovery.

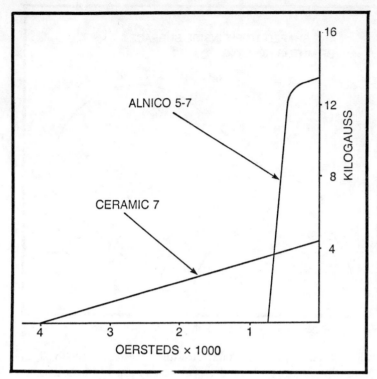

Fig. 7-9. A comparison of alnico and ceramic magnet characteristics.

In the case of the alnico magnet, the external field must be carefully limited. If the external field strength approaches the magnet's coercive strength, a drastic, permanent loss of magnetism will occur.

Ceramic magnets do have a greater temperature sensitivity than the alnico magnets do. The temperature coefficient of induction of the ceramic magnets is 0.2% per degree centigrade. This is an order of magnitude greater than that associated with the alnicos. It means that if the temperature rise in a motor application is 70°C, the magnet will lose 14% of its flux. The motor speed and torque constants will be proportionately affected.

Ceramic magnets are most commonly molded into arcs such as shown in Fig. 7-10. Being in such a shape lends to a high degree of manufacturing convenience. Magnet arcs are usually secured to the shell or housing by means of a high strength epoxy adhesive.

Because of their very high coercivity, ceramic magnets are usually magnetized while assembled into their shell. The magnets

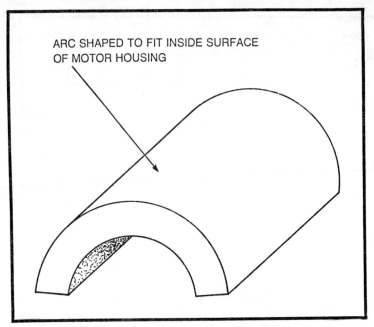

Fig. 7-10. A magnet arc typical of the shapes used in dc motors.

will then remain fully charged even with the armature removed. In order to become fully charged, it is necessary to apply to the magnet an external field strength of 20,000 amper·turns/inch. The large magnetizing current requirement is usually supplied by capacitor discharge equipment.

RARE-EARTH COBALT MAGNET MATERIALS

A third family of magnet materials has appeared as a distinct possibility for exploitation in dc machines in the near future. This family of magnet material is the rare-earth cobalts. This material is based upon elements belonging to the rare-earth group in combination with cobalt. In particular, effort has been directed at the development of samarium-cobalt.

The result is a permanent magnet with an energy product far greater than that of any other material. Samarium-cobalt materials have been produced with B_R values of up to 9000 gauss and coercivity of 8500 oersteds. The flux producing capability and resistance to demagnetization are almost twice that of the ceramic materials. Figure 7-11 compares the normal curves for samarium-cobalt with ceramic 8, the barium-ferrite with highest energy product.

Samarium-cobalt magnets require a great deal of processing to achieve the highest possible magnetic properties. The result is a very costly material that is impractical for all but the most exotic applications.

A material that is substantially reduced in cost while affording characteristics almost as good as samarium-coblat is the misch-metal-cobalt compound. Misch-metal is a mixture of the rare-earth elements and the refinement cost of obtaining samarium is thus saved.

Misch-metal magnets are produced by a powder metallurgical sintering process. The magnet pieces are pressed at high pressure in a strong orienting magnetic field. Misch-metal magnets have been produced with B_R values of over 7000 gauss and coercivity of more than 7000 oersteds.

The cost of a misch-metal-cobalt magnet is still substantially higher than a barium-ferrite magnet. However, because of the high energy product of the magnet, material cost savings are provided in other parts of a motor. A motor designed with magnets of rare-earth

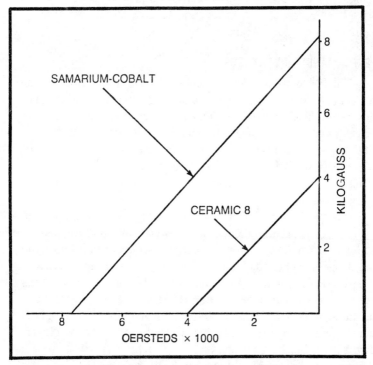

Fig. 7-11. A comparison of samarium-cobalt and ceramic 8 magnet grades.

and cobalt is characterized by compactness, light weight, and high efficiency. It can be expected that the future will see an increasing number of machines of this design.

REDUCED COMPLEXITY OF PM MOTOR CONTROLLER

Since the PM motor does not posses a field winding, it does not require the controller to provide field excitation. This is not a major advantage since most controls obtain field excitation rather simply. The field winding excitation is provided by means of a rectifier bridge circuit directly across the line voltage.

The effect, however, is to eliminate the need for several rectifying elements along with some wiring and terminal studs. Consequently a controller designed for PM motors can have several dollars of cost removed as compared to one designed for a wound-field motor.

The reliability advantage of the PM machine also evolves from this elimination of the field winding. It is a basic axiom that the fewer components there are, the fewer can fail. So it is that the PM motor drive is not susceptible to malfunction due to failure of field exciting elements in the control box.

PERFORMANCE ADVANTAGES

The PM motor provides a number of performance advantages. A chief advantage that is likely to take on even more importance in the years ahead is the inherently high operating efficiency. The efficiency of a motor can be important from two standpoints. First of all there is the intrinsic cost of the electric energy that is going to waste in a low efficiency machine. This cost of energy can be a very significant amount if a large number of integral horsepower machines are in operation.

A second viewpoint of motor efficiency may be regarded as even more important if the motor is powered by a portable battery supply. In such a case motor efficiency has a direct bearing upon the useful operating period of the battery before it must be recharged.

The PM motor has a big advantage in efficiency because it does not require a field current excitation. As will be seen later, a PM motor that is designed to fully exploit the properties of a PM field will have different proportions than the electromagnet motor of comparable rating. Similarly an analysis of machine losses will show a different division for the two machines. After all the differences are

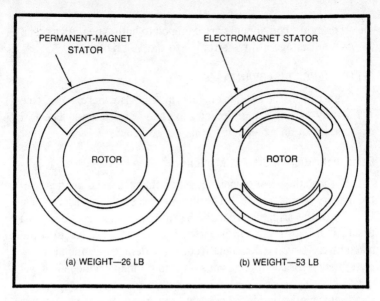

Fig. 7-12. A comparison of PM and wound-field stators from motors of the same horsepower rating.

duly noted, the difference in efficiency can usually be accounted for by the watts dissipated in the field coils of the wound-field motor.

The difference in size and weight between a PM and wound-field motor can be very dramatic. The PM motor is able, through the use of high coercivity ceramic magnets, to develop more lines of flux per pole in a given frame size. The reason for this becomes very obvious upon looking at Fig. 7-12. The PM motor utilizes two magnets each with a radial thickness of approximately ½ inch and a total weight of 3½ pounds. These four magnets provide enough flux lines to enable a 4½ inch diameter rotor to develop ¾ hp, with an active length of less than 3 inches.

Figure 7-12(b) shows a wound-field motor of similar rating. In this case the field coils take up so much space that the rotor diameter is effectively limited to 3.2 inches even though the outside frame diameter is an inch larger. Largely due to the small rotor diameter, the active length of the rotor is 4.5 inches. The end result is that the PM motor weighs 26 pounds while the wound-field (of equal rating) motor weighs 53 pounds.

It is perhaps obvious that, with the great disparity in material content, the PM motor has great advantages in achieving lower cost in manufacturing. The methods and techniques used to manufacture the other motor components are common to both types. As a result the cost advantage swings strongly to the PM motor.

PERFORMANCE DISADVANTAGES

As is usually the case, in engineering and elsewhere, any choice will usually include a mixture of good and bad features. The PM machine is not an exception to this general rule.

Demagnetization of the PM Machine

Perhaps the most obvious question that comes to mind is "can the magnets demagnetize?" The answer is yes. As was discussed in Chapter 5, the armature reaction has a demagnetizing effect. As a result, if armature current is allowed to rise to a critical level the magnets can become demagnetized. If the demagnetization is complete, the motor is out of business; without the magnetic field it will not turn.

Demagnetization is a possibility but in practice it very rarely occurs. Most PM machines have magnet proportions such that it

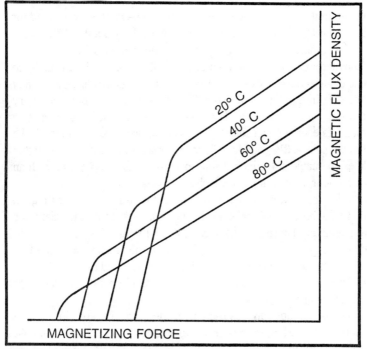

Fig. 7-13. Demagnetization characteristics as a function of temperature.

takes a current of about 10 times the full load value to obtain any measurable degree of demagnetization. The current requirement to fully demagnetize would likely be so high as to quickly burn up the armature.

A second factor also acts to minimize the possibility of demagnetization. This is the current limitation built into most controllers. The controllers are designed to provide up to 150% of full load current and then limit sharply.

There is one condition when the PM motor does become vulnerable to demagnetization. Refer to Fig. 7-13. The curves that are shown are typical of the ceramic class of permanent magnets. The curves show how the magnet material is affected by temperature. The crossover point of the curve with the horizontal axis is called the coercivity of the magnet and is an indication of its resistance to demagnetization. As can be seen by examining the curves, the coercivity reduces at low temperatures. This means that if a PM motor is allowed to soak in a low ambient temperature for a long enough period and then subjected to a large current, it is more likely to demagnetize. If a PM motor is intended for use at low ambient temperatures, special attention should be given to the duty cycle and other application conditions to determine the degree of vulnerability.

Commutation Difficulty

The PM motor is built without interpoles or compensation coils which are provided in a wound-field motor to aid commutation. As a result, PM motors will commonly show a small amount of sparking under the brushes. The sparking may be innocuous or in some cases it may pose as a major factor in unacceptably short brush life.

The point is that in the wound-field motor, the interpoles and compensating coils can overcome other design defects in the machine. In the case of the PM motor the opportunity for compensation does not exist. It is, therefore, necessary for the motor designer to be very careful to obtain the most favorable features to minimize sparking. In some applications it may also be necessary to limit the allowable time under overload during which time commutation becomes even more difficult.

COMPARISON OF WOUND-FIELD AND PM MACHINES

The PM motor is a constant strength machine. As such, it has a torque-speed characteristic that is comparable to a shunt-connected electromagnet motor. There is no PM counterpart of the series-field

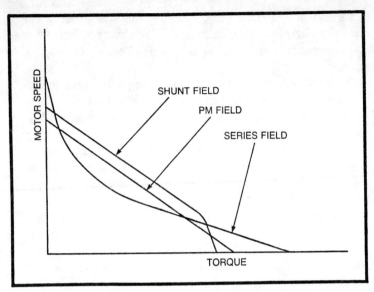

Fig. 7-14. Motor torque-speed curves for PM and wound-field types.

motor. The three motor characteristics are shown in an earlier section of this chapter; the wound field motor allots a considerable amount of space inside the frame to bulky field winding coils. Because of this loss of diameter, the wound-field motor must utilize high air-gap flux densities. Values of 50- to 60-thousand lines per square inch are typical. The air-gap flux density dictates the thickness of the rotor teeth. They must be ample enough to contain the flux lines that cross the air gap. In order to maintain the air-gap flux density as uniform as possible, the slot openings are made relatively narrow. The result is a slot design typified by Fig. 7-15. The tooth is wide, the slot relatively shallow, and the slot opening small.

The PM motor requires much less of the available space to put the magnets into. Because of this economy of space, a larger rotor can be used. The larger circumference of the rotor compensates for the low flux density of the magnet to still provide the necessary lines of flux per pole. In the PM machine the air-gap flux density is limited by the intrinsic value of the magnet material. Typical values for the air-gap density of a PM motor with ceramic magnets range from 14- to 20-thousand lines per inch square.

The much lower air-gap flux requires much less of a rotor tooth thickness to contain it. In fact very often the tooth thickness of a PM machine is decided on the basis of a minimum mechanical strength

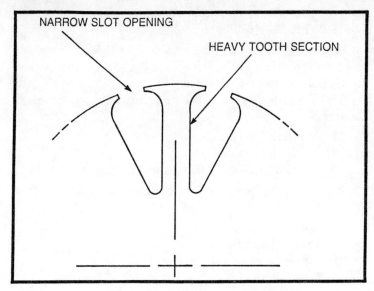

Fig. 7-15. Rotor slot geometry from a typical wound-field motor.

rather than a flux capacity criterion. Because of the high coercivity of the permanent magnet, the slot openings can be made very wide. This feature is usually an assist to producibility and also to commutation ability. The result is as pictured in Fig. 7-16. It features a large slot with a large opening and a fragile looking rotor tooth.

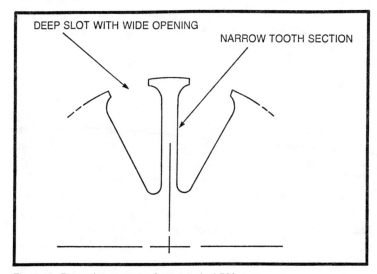

Fig. 7-16. Rotor slot geometry from a typical PM motor.

LOSSES IN THE PM MOTOR

As a consequence of the different design detail, an analysis of the power losses in a PM machine shows quite a different picture than that of the wound field motor.

The PM motor shows just three principal loss components. These are:

1. Armature I^2R loss
2. Brush contact loss
3. Rotational loss (iron, friction, and windage)

In contrast the wound-field motor shows four significant loss components. They are:

1. Armature I^2R loss.
2. Brush contact loss.
3. Rotational loss.
4. Field I^2R loss.

In comparing the PM motor losses to those of the wound-field motor, the following observations are made.

1. Armature I^2R losses tend to be lower in the PM motor. This is because, with the larger slots that can be used, larger conductors can also be accommodated. The result is less heat generation in the armature conductors.
2. Brush contact loss will be the same in both machines. This loss component is largely determined by the input voltage and current characteristics and is not greatly influenced by other design features.
3. The rotational losses are greater for the PM motor. This is because the PM motor rotor usually has a much larger rotation mass and the losses go up proportionately.
4. The wound-field motor has a significant power loss in the field winding. This power loss can usually be estimated as between 5 and 10 percent of the output power. The PM motor, because of its use of a permanent magnet field, does not have any field winding losses.

Because of this difference in power loss distribution, in particular the lower I^2R losses, the PM motor has several distinct advantages. Its temperature rise will be less sensitive to high form-factor power supplies. Consequently a PM motor designed for a battery power supply does not require as much derating when applied with rectified ac supply.

A second advantage is that less thermal stress is placed on the film insulation of the magnet wire. This is because with less power dissipating in the windings, less heat must be transferred through the insulation. Of course insulation temperature is the limiting factor in most dc motor applications.

The comparison in Table 7-1 of power losses and temperature rise profile in PM and wound-field motors of comparable ratings demonstrates the operating differences discussed in this section.

OPERATING CHARACTERISTICS OF PM MACHINES

The PM machine which utilizes ceramic magnets will show a substantial temperature sensitivity in its own voltage constant and torque constant characteristics. As the magnet temperature increases with motor warm-up, the voltage constant of the motor will decrease. This change has the effect of increasing the no-load speed of the motor.

However, as the magnets suffer a temporary loss of magnetism due to temperature rise, the slope of the speed-torque curve also increases. This effect is shown in Fig. 7-17. The crossover point of the two curves is always well to the right of the full-load torque. This means that as the motor heats up, there is a slight increase in speed under full-load torque. There is also usually a slight increase in the full-load efficiency of the motor. This is because the gain in speed and reduced iron losses will more than offset the higher current dictated by a reduced torque constant.

The motor ratings are always given for a temperature stabilized condition. Thus a motor that is rated at one horsepower will develop

Table 7-1. Comparison of Power Loss and Temperature Rise of Wound Field and PM Motors (Data Taken From 3/4 HP TEMV Motors).

	Wound Field	PM
Armature I^2R	66 watts	44 watts
Brush contact	35	32
Rotational	35	51
Field I^2R	49	-
Efficiency	75%	85%
Armature temperature rise	97°C	100°C
Commutator temperature rise	90°C	82°C
Frame temperature rise	70°C	62°C

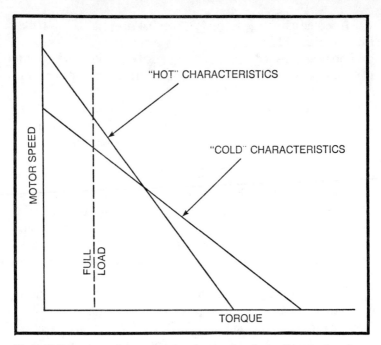

Fig. 7-17. Torque speed characteristics showing the effects of temperature rise.

one horsepower output after it is allowed to reach temperature equilibrium while loaded to full-load torque.

THE PM MOTOR AS A SERVO MOTOR

The PM motor is ideal as a servo motor in repetitive start and stop applications. The air-gap magnetic field is always there and at full strength waiting to interact with an armature current. The full response of the PM motor is obtained without power being wasted in a field winding while waiting for a command signal to go.

On the other hand, the field winding of a wound-field motor is characterized by a very high inductance. This means that after a voltage is applied to the coil an appreciable interval of time (probably hundreds of milli-seconds) is required for the field current to build to its full strength. If a wound-field motor is used in a servo application, it is therefore necessary for the field winding to remain excited at all times in order for the servo response to be of a practical value. This means a substantial waste of electric energy.

CHAPTER 8
Problems of Commutation, Sparking, and Brush Wear

The performance characteristics of the dc motor are outstanding for a wide variety of applications. Its compatibility with other circuit components in controlled systems provides an elegance of control unmatched by any other motor type.

Despite its many virtues the dc motor does have a major weakness. The weakness centers about the function of commutation. The function requires constructional features which add to the cost of the machine. In addition, the byproduct of commutation, brush sparking, will sometimes add to problems of operation and maintenance.

It is interesting to note that the problem of sparking common to modern machine design was of much lesser magnitude in the very early days of the electric industry. This was because of the early method of construction (see the Commercial Development of Electric Machines section in Chapter 1). With the very low inductance associated with the smooth cylinder armature core, commutation difficulties were unknown. As a matter of fact it was possible to use metal brushes with very low contact drop and losses associated with them.

It was not until deep slotted armature cores came into use around 1890 that excessive sparking became a problem. It was at that time carbon brushes came into use and made possible the building of larger ratings than had ever been built before.

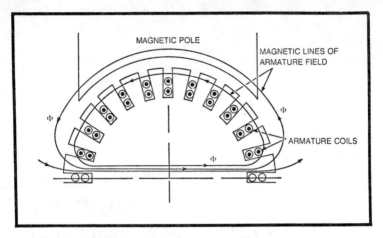

Fig. 8-1. The magnetic energy storage in the field of the armature slot.

This chapter discusses the commutation function and the causes of sparking. There are a number of different theories about, or ways of looking at, brush sparking. Consideration of these alternative views helps to an overall understanding of this problem. The factors which contribute to sparking are related to features of machine design.

ENERGY STORAGE IN AN ARMATURE COIL

In Chapter 2, the Commutation in a DC Machine section included a qualitative description of the commutator function. It was compared to the action of a rotary switch where the contacts were continuously making and breaking contact. Just as in the case of a switch when opening, there is a tendency for a dc machine to draw an arc. As the individual commutator bars pass under the brush and then break contact with it, there is a tendency to spark. The degree of sparking is a function of the machine design and can be minimized by close attention to design detail.

An important factor in determining the degree of brush sparking is the amount of energy storage associated with individual armature coils. While the primary purpose of the armature conductors and current is to create an energy conversion, there is also an element of stored energy in the magnetic field of each armature conductor. The disposition of the coil magnetic field is as shown in Fig. 8-1. The energy stored in each armature coil is given by the expression,

$$W = \tfrac{1}{2} L i^2 \tag{8.1}$$

In Eq. 8.1, the inductance L is that of the individual armature coil. The current i is the conductor current.

As the individual coils are passed under the brush the direction of current flow in the conductor changes. The change in current direction means that the magnetic energy storage must reduce quickly to zero when contact between brush and commutator bar is broken. The sudden collapse of the magnetic field returns the energy to the circuit, where it is dissipated as joule heat and if the voltage between brush and commutator is high enough, there will be a spark.

It can be appreciated that good machine design requires minimizing the coil inductances. Usually the objective of low inductance is in conflict with the objective of an efficient and economical machine. It is then a matter of careful design trade off to achieve all the desired performance levels without compromising the machine's commutating abilities.

THE CLASSICAL THEORY OF SPARKING

The very earliest dc machines were built with smooth cylindrical cores and the conductors located in a uniform air gap. This type of construction produced an armature with a very low inductance associated with it. The inductance can be viewed as an index to the level of magnetic field energy storage as described in Fig. 8-1.

These early machines used copper brushes with good success. Brush sparking as a problem had not been encountered. As the deep slotted cores came into general use, the copper brushes became unusable. It was at this time that carbon brushes came into use as a contact material, which greatly assisted in suppressing the undesirable sparking and in extreme cases, commutator flashover.

At this time a great deal of attention was given to identifying the causes for the sparking tendency. It was found that deep, narrow slots produced more sparking than shallow, wide slots. It was found that large diameter machines sparked more than small diameter machines with the same slot and winding configuration. This effect was attributed to the increased length of the end turns in the large diameter machine.

It was found that increasing the number of armature coils produced very beneficial effects on sparking. It was found that using a brush material that developed a relatively high voltage drop at the brush-commutator interface was also of great benefit.

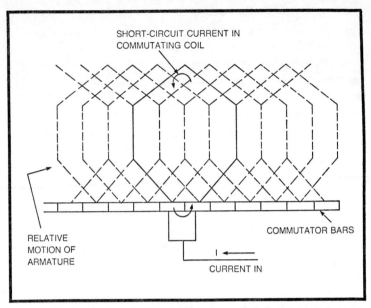

Fig. 8-2. The brush short circuiting an armature coil during commutation.

From all of the various empirical observations there gradually emerged an understanding of why the slotted core machines sparked severely. It was recognized that all of those factors that increased the sparking of the machine also served to increase the inductance, or energy storage capacity, of the armature winding.

At the moment a coil is undergoing commutation, a brush is in contact with two or more adjacent commutator bars. The brush is in effect short circuiting the coil. At that instant of time the current through the commutating coil is forced to change rapidly. The magnetic field associated with the commutating coil must also change rapidly. These forces give rise to a self-induced voltage appearing across the terminations of the short circuited coil. This voltage is called the reactance voltage (or sparking voltage) and is predicted by the relationship

$$e_R = L_C \frac{di}{dt} \qquad (8.2)$$

where L_C is the self-inductance of the single coil undergoing commutation. The differential di/dt is the time rate of change of the coil current. The reactance voltage is of a polarity that acts to continue the direction of current flow which existed in the coil before commutation.

The action of the reactance voltage can be regarded as the creation of a short-circuit current as shown in Fig. 8-2. The effectiveness of the carbon brush in reducing sparking can also be seen. The carbon brush permits the buildup of an oxide film on the commutator, which has the high electrical resistance of a semiconductor. This high resistance in the short circuit acts to absorb the energy of the collapsing magnetic field. It reduces the magnitude of the short-circuit current and thereby minimizes the resulting brush sparking.

After the causes of sparking are clearly understood, it is possible to quantify the limiting parameters. It has been found that if machines have reactance voltages in excess of 4 volts, they will be likely to have commutation difficulties.

The clear picture of reactance voltage and its effects has also allowed development of a number of special design features to neutralize it. These features will be discussed in a subsequent article of this chapter.

LAMME'S THEORY OF COMMUTATION

The classical concept of sparking voltage evolved over an extended period of time with a number of persons contributing to it. It provides a means for analyzing commutation performance and for predicting sparking.

Although an adequate theory of sparking did exist, B.G. Lamme proposed a slightly different concept which is a bit easier to visualize. The Lamme theory also results in a well organized set of equations which are straightforward in application. It is very useful to consider the Lamme theory for the greater insight it affords to the problem of sparking.

Lamme first of all develops a model for a dc machine. The model appears as in Fig. 8-3. The key features of this model are as follows:

1. There is a main field magnetizing force F which creates the air-gap flux that interacts with the armature conductors.
2. The armature conductors allow the flow of an electric current and constitute a kind of distributed coil whose axis coincides with the brush axis.
3. The armature conductors and current constitute a second magnetizing force "A" acting along the brush axis. There is a resulting distribution of magnetic flux along the brush axis. Refer to Fig. 8-4. This quadrature axis magnetic field

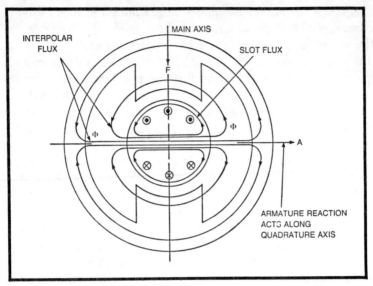

Fig. 8-3. Lamme's model for explaining his theory of reactance voltage.

is stationary and fixed in space by the function of the brushes.

4. The armature conductors rotate through the stationary quadrature axis and experience a generated voltage. The principle of this voltage generation is the same as that occurring in the region under the main field.

5. The resultant generated voltage in coils moving across the quadrature axis occurs just at the moment of commutation. If the generated voltage is great enough, the result is sparking under the brush.

The Lamme approach identifies the armature ampere·turns as the underlying cause for a reactance voltage. It identifies three fluxes which result from the armature reaction. They are: the *slot flux*, which crosses the slot from tooth to tooth; the *interpolar flux*, which leaves the armature core to pass into an adjoining iron structure; and the *end flux*, which is set up in the zone of the end turns.

In applying the Lamme theory it is first necessary to calculate a value for each of the three magnetic fluxes. The voltage due to each flux can then be calculated by considering the coil current, rotational speed, and the number of turns per coil. The total reaction voltage is equal to the sum of the three component voltages. The reactance voltage will then give rise to a short-circuit current in the commutating coil with attendant sparking. The same criteria for reactance voltage limits apply to Lamme's theory.

In calculating values for the three different fluxes, consideration must be given to the detail of slot geometry, interpolar clearances, and end-turn lengths. As it turns out, the same factors play the same role in the Lamme theory as in the classical theory. From either point of view, constructional features such as deep, narrow slots, long end-turns, large number of turns per coil, and small air gaps on the quadrature axis will cause a large reactance voltage.

It should be kept in mind that there is just one underlying cause and one effect being treated in both views of sparking. The underlying cause is the magnetic energy storage inherent in the armature coil. The result is a reactance voltage and consequent sparking. The two theories that have been presented provide different views of the same phenomena and the individual may adopt that one which he prefers.

SPARKING FROM A CURRENT DENSITY STANDPOINT

The reactance voltages that are calculated by either of the foregoing theories are average values. They are averaged over the entire commutation interval. This does not prevent the useful appli-

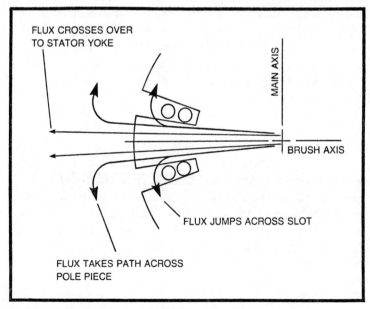

Fig. 8-4. Distribution of magnetic flux lines along the brush axis.

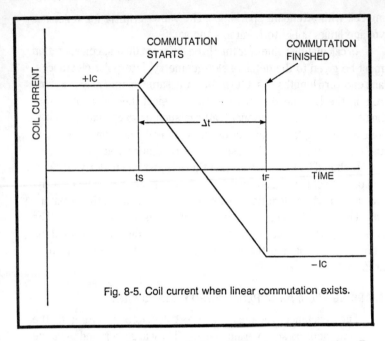

Fig. 8-5. Coil current when linear commutation exists.

cation of the theory for the prediction of effective commutation. By being average values of voltage, however, the calculations do not afford an insight into conditions that may exist instantaneously.

Additional insight into this vexing commutation problem may be obtained by viewing it from the standpoint of brush current density. Consideration of brush sparking from this perspective also provides a description of a principal mechanism of wear in both the brush and the commutator.

In the ideal situation the current density remains constant in all parts of the brush throughout a commutation cycle. If this condition is achieved, the commutation is said to be linear. The current in the commutating coil will change as shown in Fig. 8-5. It is seen that prior to commutation commencing at t_S, the coil current is $+I_C$. After commutation commences and during the interval Δt, the current falls along a straight line until at t_F the value of coil current is at $-I_C$. It has the same magnitude of current but the opposite polarity as at t_S. When linear commutation is achieved the (di/dt) of Eq. 8.2 is a constant value. Then,

$$\frac{di}{dt} = \frac{\Delta i}{\Delta t} = \frac{2I_C}{\Delta t} \qquad (8.3)$$

When the ideal linear commutation is achieved, the maximum induced reactance voltage will be equal to the average value. If the calculated reactance voltage is less than the sparking criterion there will be good commutation.

If the current in the commutating coil does not change at a linear rate it might appear as in Fig. 8-6. In this instance the rate of the current change is delayed after commutation commences at t_S. The coil current continues to flow towards the leaving commutator bar. This means that the area of contact between the leaving bar and the brush is decreasing at a higher rate than the current. The result is that the current density between the brush and leaving bar reaches very high levels. At t_F, when the brush and bar are about to break contact, there is a tremendous concentration of energy at the leaving edge of the commutator bar. This concentration of energy may be enough to heat the copper to its melting point. The result is the vicious spark associated with poor commutation.

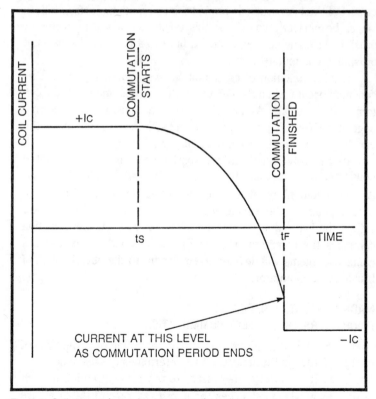

Fig. 8-6. Coil current for undercommutated condition.

If the concentration of energy due to high current density is sufficient to cause an instantaneous heating sufficient to melt the copper there will be an orange spark. This is a sign that a harmful condition exists that is likely to get worse. If the concentration of energy is not sufficient to cause melting of the copper there may still be a spark, but the spark will be silvery or bluish in appearance and is relatively innocuous.

When an orange colored spark does exist it produces conditions that additionally compound the problem. The result is a steadily increasing severity of sparking which will deteriorate the commutator and brush at an increasingly fast rate.

The mechanism for this progressive worsening of commutation works in this way. As copper is melted at the leaving edge of the bar it is removed by the resulting spark. As a consequence the commutator contact area is slowly eroded away. This causes the current density to reach ever higher values at the moment of breaking contact. This causes greater heating and more sparking. As a greater part of the bar surface is eroded the commutator surface will no longer be perfectly round. At this point mechanical problems of brush bounce and chatter will contribute to a very rapid erosion of brush and commutator.

A third condition of commutation is shown in Fig. 8-7. This condition occurs when the current in the commutating coil reduces to zero very quickly. As a result the current density in the entering edge of the bar reaches very high levels with resulting melting of the bar. This condition is called overcommutation.

It is usually difficult to see sparking for an overcommutated condition. This is because the sparking is occurring under the brush. As the condition continues, though, copper is displaced from the entering edge to the leaving edge. This causes a high spot on one side of the bar and a low spot on the other side. As the condition continues the commutator becomes out-of-round and mechanical contact problems will develop to contribute to the deterioration of brush and commutator.

MACHINE ELECTRICAL DESIGN
PARAMETERS THAT AFFECT COMMUTATION

As has been mentioned in a previous section, it is the level of energy storage in the armature that determines the reactance voltage level. It is therefore necessary to pay close attention to such design detail which influences the energy storage of the armature

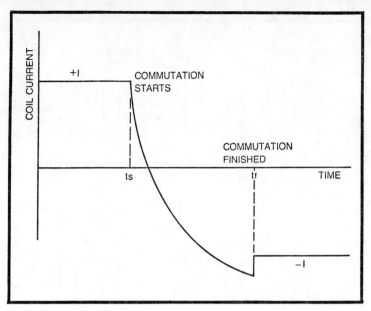

Fig. 8-7. Coil current in an overcommutated machine.

windings. This design requirement will often dictate that a machine not be designed in the most economical way from a strictly performance standpoint. If the machine designer did not have to worry about the commutation aspect he would be primarily concerned with the thermal capacity of the machine. This design approach allows an economy of material utilization that is not usually achieved when commutation limitations must be considered.

As a concession to better commutation, many dc machines are designed with shallower slots than the maximum possible in a given geometry. For the same reason, the slot opening at the air gap will also be designed with a wide opening. This design consideration is illustrated in Fig. 8-8.

Another important design consideration concerns the geometry of the interpolar space. In particular the arc of the main pole face plays an important role in determining the level of the reactance voltage. A design with relatively narrow main pole faces will be more effective in commutation than one that has a very wide arc. The narrow pole face reduces the magnetic field in the interpolar gap which is created from armature reaction. The lower level of magnetic field means a smaller reactance voltage in the commutating coil. This design principle is illustrated in Fig. 8-9.

183

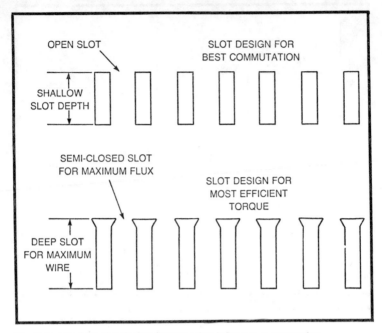

Fig. 8-8. Armature slot design features that influence commutation.

The most effective way of minimizing the reactance voltage is to limit the number of armature windings that are commutated at a time. This means that the armature must be designed with fewer turns per coil and a larger number of coils per armature. While this design approach is effective in reducing sparking voltage it unfortu-

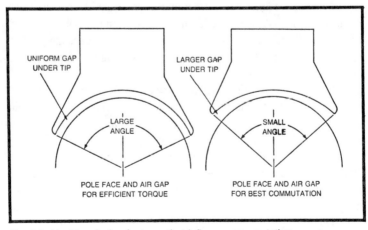

Fig. 8-9. Machine design features that influence commutation.

nately also acts to increase manufacturing costs. More armature coils means a more expensive commutator plus the added cost of making the additional electrical connections. Figure 8-10 illustrates this design principle of increasing the number of coils to decrease the reactance voltage.

MACHINE FEATURES THAT NEUTRALIZE ARMATURE REACTION

There are a number of features that can be designed into a machine which are very effective in neutralizing the causes of poor commutation performance. These design features include:

- Moving the brush axis off neutral.
- Providing a special commutating pole (also called interpole).
- The use of compensating windings in the stator assembly.

The underlying theory for all of these design features is the same. It goes like this:

- The cause of sparking in a commutating coil is because of a magnetic field effect.
- Introduce another magnetic field source which will oppose the effect causing sparking.
- The result is a machine with no reactance voltage.

Depending upon the difficulty of commutation in a particular machine, one or all of the methods may be used. Again, as so often is the case, incorporation of features to improve commutation will invariably increase manufacturing costs.

MOVING THE BRUSH AXIS OFF NEUTRAL

Figure 8-3 illustrates the orientation of the main field and the armature reaction where the brush axis is on the neutral. The

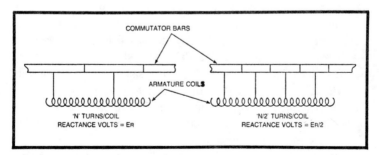

Fig. 8-10. Increasing the number of commutator bars decreases reactance voltage.

185

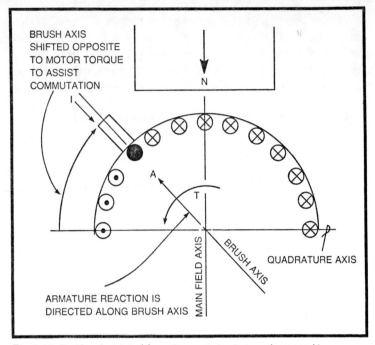

Fig. 8-11. Brush axis rotated from the neutral axis to reduce sparking.

neutral axis is at quadrature with the main field axis. With this orientation a reactance voltage is generated in the armature coils as they sweep past the armature reaction axis. With this arrangement the main field and armature reaction are in space quadrature and have no interaction or effect on one another.

In Fig. 8-11 is illustrated an arrangement with the brush axis rotated away from the neutral axis. It can be seen that with this orientation a component of armature reaction is in opposition to the main field. Since the reaction voltage in the commutating coil is due to armature reaction, the idea is to rotate the brush axis until armature reaction is equally offset by the main field. If this condition is achieved, commutation is greatly improved.

In order to utilize brush shift as an assist to commutation, the brushes must be turned so that the main field and armature reaction will oppose. Another way of stating it is that the brushes must be shifted in a direction opposite to the rotation for motors and in the direction of rotation for generators.

Brush axis shifting is effective in helping commutation in many instances, but it does have definite limitations. Perhaps the severest limitation is that a brush shift will assist only for one direction of rotation. If the machine is operated in the wrong rotation, commutation will be severely worsened. For this reason this stratagem is used only with unidirectional applications of dc machines.

Another drawback of brush shifting is that in the opposition of main field and armature reaction, some air-gap flux is also lost. This means the machine is somewhat weaker and will decrease the margin of machine stability.

COMMUTATING POLES (INTERPOLES)

The disposition of a commutating pole in a dc machine is illustrated in Fig. 8-12. In this arrangement the brush axis is maintained in quadrature to the main field axis and there is thus no interaction between the main field and the armature reaction.

The interpole is a special pole centered on the neutral axis. Its winding is connected so that it carries the same current flowing through the armature. The number of turns on the interpole is selected so that it offsets the effects of armature reaction.

The interpole will cover just a small part of the air gap, in contrast to the main pole design. Its purpose is to introduce its

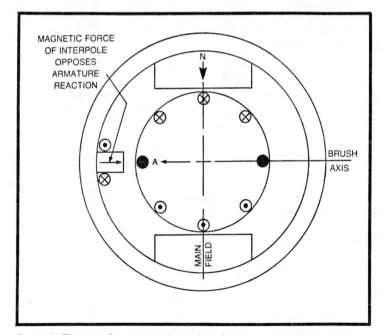

Fig. 8-12. The use of a commutating pole to improve commutation.

Fig. 8-13. Special laminated stator structure for minimizing commutation problems of rectified power. (Courtesy of Reliance Electric Co.)

neutralizing effect just during that period of time required to commutate the coil passing under it.

The interpole feature is far more effective and flexible than brush shifting in assisting commutation. By virtue of its excitation by the armature current it automatically adjusts the interpole field strength to the proper level for the armature load. It can also be used with machines used in bidirectional applications as long as the proper polarity is maintained between the interpole and armature windings.

A special problem with interpoles is sometimes encountered when motors are used with rectified ac power. In such cases it is possible, because of the pulsating nature of the armature current, for eddy currents in the motor shell to significantly reduce the effectiveness of the commutating poles. The eddy current opposes the establishment of the interpole flux and delays it, causing a phase shift between the flux and the armature reaction. In such instances it is necessary to use a stator with a laminated core structure. Figure 8-13 shows the construction features found effective for use with rectified power sources.

COMPENSATING WINDINGS

Compensating windings are used in the same way and for the same purpose as the commutating poles. They produce a magnetic field in space quadrature to the main field. They are provided with a polarity to oppose armature reaction and are designed with a strength to offset the undesirable effects originating from the armature.

The arrangement of compensating windings is as shown in Fig. 8-14. The compensating coils are located in slots in the face of the main pole. The windings are connected so as to carry the armature winding and thus automatically adjust to the right strength.

The use of compensating windings also allows for bidirectional application of the machine. Current flow through the compensating winding is merely reversed as the armature voltage changes.

Compensating windings also provide a means of counteracting the demagnetizing effects of armature reaction. In contrast to interpoles, whose effects are concentrated on the neutral axis, the compensating winding is distributed around the air gap. This provides a means of machine performance improvement beyond that afforded by interpoles.

MECHANICAL FACTORS OF DESIGN AFFECTING BRUSH SPARKING

As has been described in the preceding sections, the arrangement of the magnetic fields and electric conductors plays an important role in whether a machine will spark or will commute successfully. In addition to the electromagnetic causes there are also a

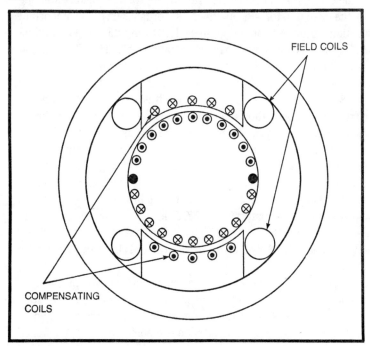

Fig. 8-14. Compensating windings for improving commutation.

number of possible design causes which are more of a mechanical nature.

The mechanical causes of sparking include such things as: proper brush material, adequate brush area to handle armature load currents, a well balanced rotor, a true commutator surface with good surface finish, and a brush holder or support system which maintains intimate contact with the commutator at all times. If a machine is deficient in any of these areas it is possible for severe sparking to occur even though the reactance voltage is at a level well below the sparking criterion.

For the reason mentioned above it is vitally important that a great deal of attention be given to these more mundane design features. After a machine design is finally regarded as being optimized it is usual for the manufacturer to enter extended periods of testing in order to find the proper grade of brush and to assure the adequacy of the brush support.

THE FUNCTION OF THE BRUSH IN COMMUTATION

The primary function of the brush in the dc machine is that of a sliding contact. It transfers electric current to the armature winding via the commutator. In addition, the brush serves an important function during the commutation cycle. During this interval of time the brush short circuits the commutating coil. The short circuit which flows during this period of time is limited only by the resistance of the short circuited coil and the contact resistances (refer to Fig. 8-2). Since the coil resistance itself is very low, the principal current limitation is then the contact resistance between the brush and commutator. In larger machines the dependence upon the contact resistance is even more pronounced since the coil resistance becomes negligible. The function of the brush during commutation is to absorb and dissipate the stored energy of the coil involved.

Because of the varied demands on the brush, the use of carbon and graphite as a material has become general for dc machines. Carbon provides the required high contact resistance with an ability to withstand extremely high temperatures and also provide a degree of self-lubricity.

In most cases where there is an apparent rapid wearing away of carbon brushes it is due to mechanical abrasion of the commutator. The chain of events is that the commutator surface is damaged by electrical effects during commutation to a point where its roughened surface causes rapid mechanical wear of the brushes.

An idea of the relative abilities of brushes and commutator to withstand heat may be obtained by considering their limiting temperatures. The melting point of copper is 1083°C. On the other hand carbon does not exist in a liquid state but is transformed from solid to vapor at temperatures between 3600°C and 4000°C.

FILM FORMATION BY CARBON AND GRAPHITE BRUSHES

In operation, carbon brushes are made effective by developing a film on the commutator surface. The film has been found to be composed of copper oxides mixed with brush materials. The combination of brush and commutator film also provides a high degree of lubricity. The lubricity is attributed to a layer of water molecules which adhere to the surface of the brush at the commutator interface. The source of the water molecules is the surrounding ambient air at normal conditions. A condition which reduces the water content of the air, such as high elevations and cold and dry environments, will effect the formation of the water film. The result will be a great increase in friction with consequent heating of the surfaces and rapid wear of brushes and commutator.

Film development results from the deposition of brush material onto the commutator surface by the negative brush. While the material deposition is occurring, copper is also oxidizing at the commutor surface under the temperature conditions which exist there. The positive brush acts to polish the commutator surface by virtue of the friction between the two parts. The film thickness can then be seen to result from a combination of current density, brush material characteristics, peripheral speed, and brush pressure.

The thickness of the film determines the color of the patina and also influences the contact drop between brush and commutator. The heavier the film, the darker will be its appearance. On the other hand a film that is only microns thick will be of a very light shading. The color of the commutator patina is not always a good indicator of commutation effectiveness nor is too thick a film necessarily good.

A thick film on the commutator will provide the following desirable effects:

- low friction.
- high contact drop.
- minimum commutator wear.

The bad effects from too thick a film can be listed as:

- Copper picking and streaking.
- High commutator temperatures.

- Irregularity in friction causing brush chatter.
- Sparking and shortened brush life.

TYPES OF CARBON BRUSHES

There are four main classes of brush materials used in dc machines. Each of the main types has very distinctive properties and characteristics which make it most appropriate for different applications. The most commonly used grades are:

1. Natural graphite. This material is used on machines where high commutator surface speeds are encountered. It is the brush type with the highest degree of lubricity and also provides a high degree of mechanical damping. It is characterized by good thermal conductivity and a high contact drop. It is suitable for use on high voltage machines.

2. Hard carbon. This material has superior mechanical qualities of strength and toughness. It is also characterized by low thermal conductivity and electrical conductivity. Its use must be limited to relatively low surface speeds and low current density.

3. Electrographite. This material is produced by exposing carbon to temperatures in excess of 2500°C. The material so produced combines some of the properties of carbon and natural graphite. It has high contact drop, good lubricity, and good thermal conductivity. It also has high mechanical strength and toughness. Electrographites are commonly used in difficult to commutate machines.

4. Metal graphites. This material is made by incorporating a metal (such as copper or silver) into an electrographitic material. The result is a brush with very low contact drop, good thermal conductivity, and very large current capacity. Metal graphites are used on low voltage, high current devices where commutation is not a problem. The percentage of metal in the brush can be varied 10% to as high as 90%, depending upon the characteristics desired.

   Within each of the main groups of brush materials is a wide range of characteristics. It is also possible to obtain special treatments for brushes intended for use in unusual environments. Motors intended for use at high altitude or in vacuum may have brushes impregnated with an oil or a special dry film lubricant such as moly-disulfide.

Chapter 9
The Application of DC Motors

The dc machine reached its zenith in the last decades of the nineteenth century. Then, with the invention of the induction motor and the transformer, power systems became standardized on 60 hertz ac. With the adoption of an ac system, dc machines fell into decline and were used only in very special applications where the unique characteristics of a dc drive system justified the much higher cost.

During this period of time, which extended from the turn of the century up to the middle of the 1960s, the most popular form of dc speed control was the Ward Leonard method. A schematic of a Ward Leonard system is shown in Fig. 9-1. The main components consisted of an induction motor which ran at essentially constant speed directly off the ac line, a dc generator, and a dc motor. The Ward Leonard system provided a high degree of control to a drive that was continuously adjustable and very smooth in operation. The obvious disadvantage of the system is the high initial cost due to the three major components.

Over the years industry has continued to have a great need for adjustable speed drives. The need is created from a diversity of automatic processing equipment which can increase productivity and justify capital investment. Because of the needs of industry, strong efforts were made over the years to utilize dc motors in combination with the 60 hertz electrical supply. These efforts took the form of rectifying devices with some feature that made them controllable.

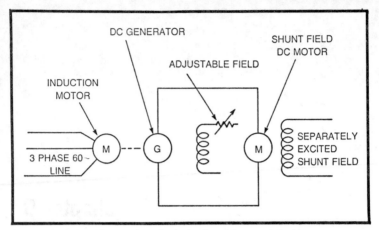

Fig. 9-1. The Ward Leonard method of adjustable speed dc drive.

Examples of such controlled rectifiers are the ignitron, excitron, and thyratron. Unfortunately none of these devices was completely satisfactory in providing the desired overall system performance. As a consequence dc drivers continued to be used only in the more discriminating applications.

The advent of solid state electronics has opened a new era in dc motor application. Starting in the late 1950s, solid state controlled rectifiers became available as commercial devices. These devices, SCRs and thyristors, provide a means for interfacing dc motors to the ac line with a degree of control and at a cost that is superior to any other sytem. As a consequence the most commonly used controlled speed industrial drive is a combination of dc motor and a thyristor converter.

The type of motor that is used in adjustable speed drives is the shunt-field motor. In recent years PM motors have made their appearance in the smaller sizes up to several horsepower. Either type of motor has the same linear torque-speed characteristic that makes it readily adaptable for speed control.

While the vast majority of fixed installation motors use rectified 60 hertz power there is also a large area of application utilizing "pure" dc. The combination of storage battery and dc motor has always been a logical choice for mobile applications. Such applications include industrial lift trucks, golf carts, recreational vehicles, marine craft, and many others.

The advent of power electronic devices has had just as great an impact on battery powered motors as those driven by controlled rectifiers. Power transistors and SCRs consititute a solid state switch that can be used very advantageously to control the dc from a battery supply.

Rectified Alternating-Current Power

An alternating current power source is converted into a dc source by means of a diode. A simple circuit schematic illustrating the function of a diode is shown in Fig. 9-2(a). The diode is a device that has a very low impedance to electric current flow for one polarity, but a very high impedance for the opposite polarity.

The diode conducts current during the positive half cycle of the alternating voltage. It blocks current flow during the negative half cycle of the alternating voltage. The resulting voltage and current waveforms are as shown in Fig. 9-2(b). It can be seen that for this particular circuit, the electric power flow to load resistor R is a series of current pulsations. The instantaneous value of current is a function of the applied voltage and the load resistance.

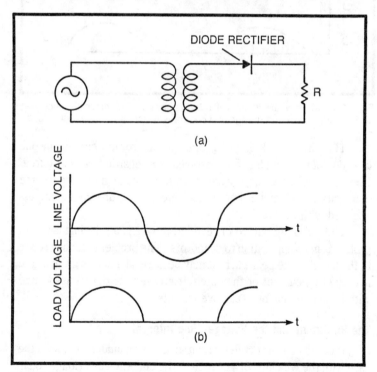

Fig. 9-2. The use of a diode permits an ac power line to be used to power dc motors.

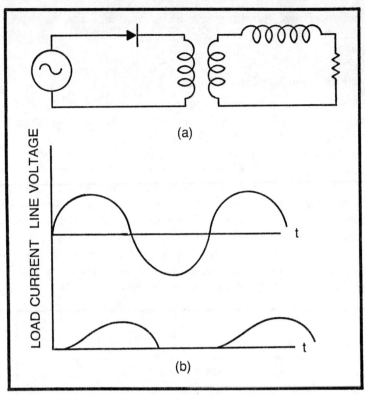

Fig. 9-3. At (a) the circuit diagram of a single phase, half-wave rectifier supplying an RL load, and (b) the voltage and current waveforms as seen by the load.

The circuit of Fig. 9-2 produces an average current in one direction of flow and for this reason can be regarded as a dc source. However, the time varying nature of the voltage gives rise to time constants and complex impedance effects that are commonly associated with ac circuits.

The pulsating nature of rectified power is a principal cause of problems in its application to dc motors. The problems usually relate to the fact that the peak current may be several times as great as the average (dc) current or they stem from the time delays associated with inductive and/or capacitive circuits.

The RL Circuit and the Voltage-Time Integral

If the load on a rectifier circuit includes an inductance, as is the case with the windings of a dc motor, the current will no longer be in phase with the applied voltage. In this case the inductor absorbs, or stores, energy in the magnetic field associated with it. During this period, when the applied voltage is being absorbed, the voltage that is induced in the coils resists the establishment of the rectified current.

The result is shown in Fig. 9-3. The current wave form is delayed in reaching its peak value. Then, even after the voltage has reversed polarity, current will continue to flow in the forward direction. The current that continued to flow after the polarity has reversed results from the energy storage of the circuit inductance. As the current through the coils decreases, the magentic field it established collapses and returns the stored energy to the circuit in the form of a current.

The current response in an inductive circuit fed with rectified power can be regarded as having two components. The steady state component is given by

$$I = \frac{e}{Z} \tag{9.1}$$

where $Z = \sqrt{R^2 + X^2}$

The second component of current response is a transient current given by the expression

$$I = \frac{e \sin \theta}{Z \ \epsilon^{Rt/L}}$$

where $\theta = \tan^{-1}$

$$\frac{wL}{R} \tag{9.2}$$

The response current waveform with its components is shown in Fig. 9-4. The ratio L/R is recognized as the familiar electrical time constant of the circuit. It can be seen that for a given value of resistance, a circuit with higher values of inductance will sustain current over a longer period of time.

Figure 9-5 shows the voltage-time integral criterion for current flow. This criterion has as its basis the expression for inductance.

$$L = \frac{e}{di/dt}$$

The terms are rearranged to give the equation

$$Ldi = edt$$

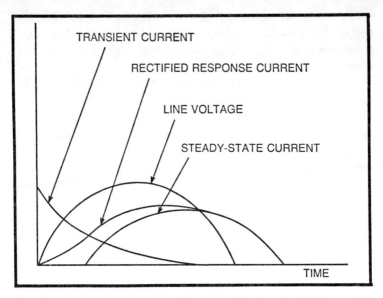

TRANSIENT CURRENT

RECTIFIED RESPONSE CURRENT

LINE VOLTAGE

STEADY-STATE CURRENT

TIME

Fig. 9-4. Component current waveforms in an inductive circuit fed by rectified power.

If both sides are integrated

$$\int L \, di = \int e \, dt$$

The term $\int L \, di$ represents the total energy storage associated with the current while varying between its maximum and minimum values. The time integral $\int L \, di$ is represented by the single cross-hatched area of Fig. 9-5. The second integral, $\int e \, dt$, is represented by the cross-hatched area of Fig. 9-5. It represents the potential energy stored by the inductor due to the absorbed voltage during the buildup of the current. The equal area criterion is a means of showing graphically that, when energy conversion is not involved, all energy stored in a circuit is returned at a change of state.

The irregular wave shape of the rectified current that is actually supplied to a load has given rise to a figure of merit to describe it. The figure of merit is called the form factor. Form factor is the ratio of rms current to the average (dc) current. It should be recalled that rms current is a measure of its heating effect. On the other hand the useful output of a device such as a dc motor is determined by the average input current. The form factor of a rectifier and load can then be interpreted as an index to the heat generation per unit of current.

Rectifier and motor combinations that are characterized by a high form factor will likely produce a hot motor if operated with high form-factor current. Unless the motor has been designed to operate at high form factor it may be necessary to derate its torque capability.

Controlled Rectifiers

From Eqs. 9.1 and 9.2 it can be seen that the current response in a rectifier circuit is a function of the applied ac voltage and the complex circuit impedance. These expressions might well lead one to deduce that control of the ac line voltage affords the best means of controlling the power in a rectifier circuit. Such is not the case. Rather, use is made of the fact that rectifier devices are available that will not conduct in the forward mode unless an electric signal is first applied to a third terminal or gate. (In contrast a simple diode is a two-terminal device.) Such a device is shown schematically in Fig. 9-6.

A characteristic of a controlled rectifier, such as that shown in Fig. 9-6, is that once it is triggered onto the conducting mode, the device will stay on until the current falls to zero. Of course when used with an ac power source, the current through the device falls to zero at the end of each half cycle when the applied voltage reverses

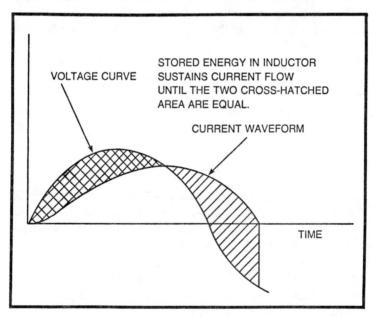

Fig. 9-5. The voltage-time integral method for determining total conduction time in an inductive circuit.

199

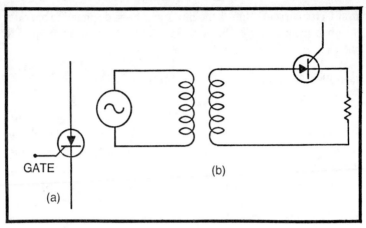

Fig. 9-6. At (a), the schematic of a controlled rectifier, and (b) a controlled rectifier used in a power supply circuit.

polarity. The effective dc output of a controlled rectifier is then controlled by adjusting the instant of time during the positive half cycle when the gate is triggered on. The voltage and current waveforms of such a controlled rectifier are shown in Fig. 9-7.

The method illustrated in Fig. 9-7 is known as phase control and is the type of control most commonly used with thyristor devices. The effective dc output of the device is regulated by controlling the time interval during which current is conducted. This in turn is a function of the angle of the voltage wave at which triggering occurs. The phase angle at which the thyristor is triggered is commonly called the firing angle.

If the load requirement is such that only a small part of the available voltage is needed, the firing angle will approach 180°. However, if the load requires the full potential of the voltage, the firing angle may approach 0°.

The rectifier device is triggered by an electric signal taken directly from the line. Line voltage is used so that the trigger signal is always in perfect synchronism with the voltage applied to the load. The trigger circuitry of a phase controlled rectifier will condition the signal to be applied to the gate of the thyristor. It will be designed to provide a signal voltage well above the threshold level necessary to initiate conduction. It will usually also shape it into a pulse with a duration short enough to allow precise control but long enough to assure that rectifier conduction is sustained.

A simple firing circuit is shown in Fig. 9-8. In this circuit the zener diode acts to limit the voltage across the unijunction transis-

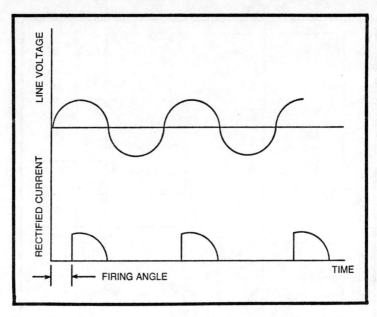

Fig. 9-7. Voltage and current waveforms with phase controlled rectification.

tor. A current charges the capacitor until the capacitor voltage exceeds the breakdown voltage of the transistor. When breakdown occurs the transistor conducts current through the pulse transformer which in turn triggers the gate of a thyristor in the rectifier power stage.

The firing angle of the thyristor is determined by the time required to charge the capacitor. The capacitor charging rate, in

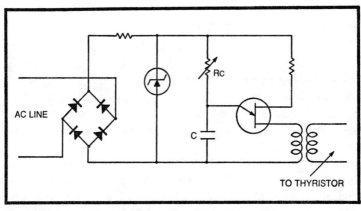

Fig. 9-8. A typical firing circuit of a phase controlled rectifier.

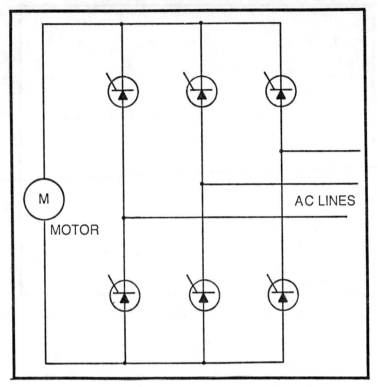

Fig. 9-9. Power supply C. Three-phase, full-wave, full-control rectification.

turn, is controlled by the RC time constant of the trigger circuit. The firing angle is varied by adjusting the value of the control resistance R_c, which is a potentiometer.

Types of Rectified D-C Supplies

The discussion in the previous sections of this chapter dealt with a single-phase half-wave rectifier. This configuration is the simplest rectifier design, but is by no means the only one in common usage. Perhaps the most common rectifier design is the single-phase full-wave circuit. This rectifier circuit is the one most commonly used with small integral and fractional horsepower motors in adjustable speed drives.

A NEMA standard has been developed to categorize rectified power supplies according to the number of current pulses per second from a 60 hertz supply. According to the NEMA standard, the rectified power supplies are defined as follows:

- Power supply C (see Fig. 9-9) designates a three-phase full-wave power supply with six total pulses per cycle and six controlled pulses per cycle. The designation applies to a supply with no free wheeling, with 60 hertz input, and without series inductance added to the armature.
- Power supply D (see Fig. 9-10) designates a three-phase semi-bridge having three controlled pulses per cycle with freewheeling, with 60 hertz input, and without series inductance being added to the armature circuit.
- Power supply E (see Fig. 9-11) designates a three-phase single-wave power supply having three controlled pulses

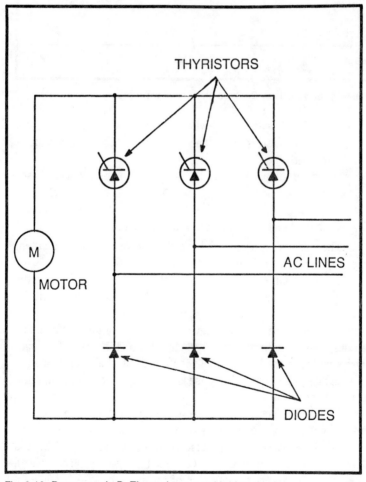

Fig. 9-10. Power supply D. Three-phase, semi-bridge circuit.

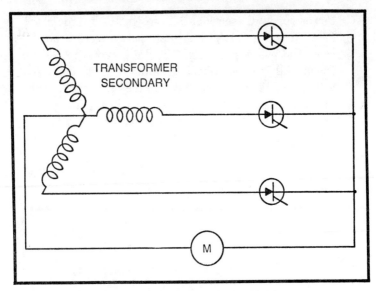

Fig. 9-11. Power supply E. Three-phase, half-wave circuit.

per cycle without freewheeling, with 60 hertz supply, and with no series inductance being added to the armature circuit.

- Power supply K (see Fig. 9-12) designates a single-phase, full-wave power supply. It has two pulses per cycle with freewheeling, with 60 hertz input, and with no series inductance being added to the armature circuit.

The various supplies each have characteristics that make them most appropriate, depending upon the application.

Supply C is the most expensive of the four types. Its greater cost is due to the six control devices along with the attendant trigger circuits etc. This circuit arrangement requires the simultaneous firing of two of the control devices for current flow. When this occurs, line-to-line voltage is connected directly across the load. This supply has the lowest ripple content of all. It also provides the greatest degree of control. Its use is usually limited to large integral horsepower motors.

Supply D (also called a three-phase half-control supply) uses a free-wheeling diode connected across the load to sustain current flow. This supply is a compromise between designs C and E. It is less costly than the full-control design, having only three control elements. However, the number of pulses per cycle becomes a function of the firing angle. For small firing angles the supply shows six pulses per cycle and a ripple content similar to full control. At larger firing angles, though, the supply shows three pulses per cycle and more closely approximates the characteristics of the three-phase half-wave design, supply E.

Supply E is the most economical of the three-phase supplies. It has only three control devices, but it does require a neutral connection. Each of the control devices conducts for 120° during each half cycle and current returns via the neutral connection.

Supply K is the most economical of all the supplies. It has only two control devices and the freewheeling diode. It also has the advantage of not requiring a three-phase line to be available. However, the single-phase supply also represents the most difficult application for a dc motor. The harmonic content of the supply current is quite high; this in turn causes difficulty of motor commutation and overheating. Special precaution must be taken in matching the motor to a single-phase supply.

Methods of Motor-Speed Control

Previous portions of this book have referred to the simple elegance of speed control possible with a dc motor. In its simplest form, effective speed control is obtained merely by inserting a variable resistance into the armature circuit of a shunt-field motor.

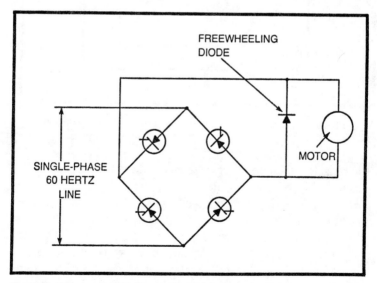

Fig. 9-12. Power supply K. Single-phase, full-wave circuit.

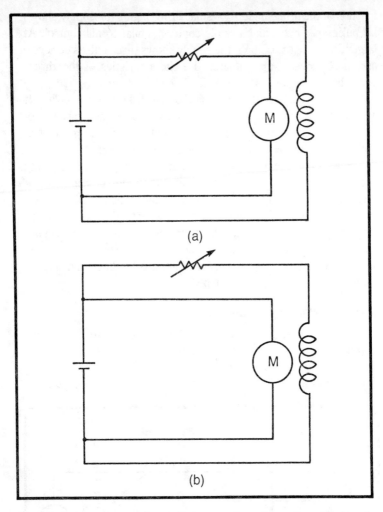

Fig. 9-13. At (a), motor speed control by means of armature voltage adjustment, and (b) motor speed control by means of field current adjustment.

The variable resistance effectively reduces the voltage applied to the armature. The reduced voltage causes a proportionate reduction in the motor speed. This type of speed control by means of a variable armature circuit resistance is shown in Fig. 9-13(a).

A second means of adjusting motor speed is with a variable resistance in the field circuit. Such an arrangement is shown schematically in Fig. 9-13(b). In this arrangement the increased resistance causes a reduction in exciting current. The reduced excitation means less air-gap flux with the result that the motor must run faster for a given value of voltage applied to the armature.

Generally, field control is not used where the motor is required to run at loads close to its rated torque. This is because the reduced field strength can cause less favorable commutating conditions and also increase the required armature current. On the other hand, field control is very useful in extending the high speed range of a motor under light load. As compared to armature voltage control, the field rheostat is required to handle relatively low power levels.

Motor speed control by means of variable resistance in the motor circuits is not a desirable method. In both cases, but especially in the case of the armature resistor, the resistances mean a substantial loss of power that is dissipated as heat. For this reason the advent of solid state switching devices such as controlled rectifiers was an important development. It was thus inevitable that controlled rectifiers, with the continuously adjustable dc voltage they provided, would be utilized in dc motor drives.

The SCR Adjustable-Speed Motor Drive

An adjustable-speed drive system combines a motor, a controlled rectifier supply, and an assemblage of control circuitry to provide the desired function. The main components of the drive system are shown in the block diagram of Fig. 9-14.

The drive system may use single-phase or poly-phase power, the function being the same with either supply.

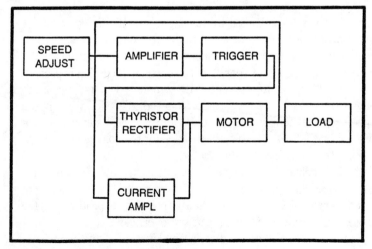

Fig. 9-14. An adjustable-speed drive system.

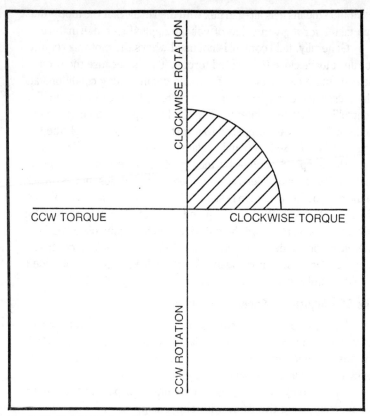

Fig. 9-15. Characteristics of a regulator type of speed control.

The speed in adjusted by turning a potentiometer, which in turn provides a reference voltage to the speed summing node. The other inputs to the speed summing node are the voltage across the motor (reflecting motor speed) and a second voltage that is proportional to the armature current. The armature voltage signal is a negative feedback which subtracts from the reference signal of the speed adjust potentiometer. The armature current signal is a positive feedback that makes a compensation for the voltage drop due to the internal impedance of the motor.

The amplifier is usually equipped with a current limiting feature. This current limit is usually designed for about 125 to 150 percent of rated current. The current limitation is necessary to protect both the motor and the SCRs from destructive overheating.

The output of the speed error amplifier is fed into the trigger circuit. The firing angle of the controlled rectifiers is then either advanced or delayed according to the input of the amplifier. The trigger circuit output is kept synchronized with the alternating voltage applied to the controlled rectifiers by the synchronizing circuit, taking power form the same line.

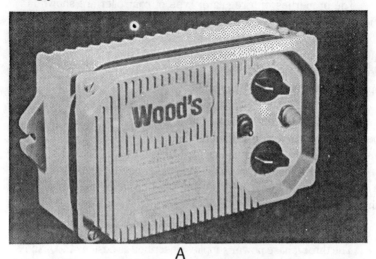

A

B

Fig. 9-16. A 1 hp motor controller, (a), a view of the control closed, and (b) an inside view of the control. (Courtesy T.B. Woods and Sons Co.)

A drive system such as shown in Fig. 9-14 is not a true servo system in that the speed of the motor shaft is not actually measured, and torque is developed in only one direction. Rather it is of a class known as a single-quadrant controller.

The properties of a regulator type of controller are shown in Fig. 9-15. It is seen that torque is developed in only one direction of rotation and speed is regulated only in that same direction. A single-quadrant controller can be operated in the opposite direction. This is done by throwing a switch which reverses the polarity of the voltage applied to the motor field winding.

The single-quadrant controller is a very effective solution for most controlled speed motor requirements. It is possible to compensate the controller so as to obtain speed regulation of 2 percent of set-speed from no-load to full-load. It is possible to obtain such speed regulation over a speed range of 50:1 of base speed. Fig. 9-16 shows a commercial motor controller used with a 1 hp motor and illustrates the compactness of design.

Operating Characteristics of SCR Drives

The three-phase power supplies described as designs C, D, and E in the Types of Rectified DC Supplies section are characterized by an inherently low ripple in their output. In addition, the harmonic frequency present in the voltage applied to the motor is high enough to be effectively attenuated by the inductance of the load. The net result is that three-phase rectified power supplies have a very good form factor associated with them. Three-phase rectified supplies will typically exhibit form-factor values of 1.05. This low form factor is a decided advantage in the application of the motor.

The situation with single-phase power is much less favorable. Figure 9-17(a) shows a typical voltage and current waveform from a ¾ hp system with free wheeling diode as shown in Fig. 9-17(b). The voltage waveform shows a clipped appearance up to the firing angle, α. The flat voltage represents the generated voltage of the armature. At α, the SCR is fired and the line voltage appears across the motor. As the line voltage reaches zero the armature current continues to flow through the freewheeling diode for an interval of time until the energy stored in the motor's windings is returned to the circuit.

The voltage waveform of Fig. 9-18(a) reveals that a minimum firing angle of about 45° is required in order to initiate rectifier current. At angles less than this the generated voltage of the motor exceeds the line voltage and current will not flow.

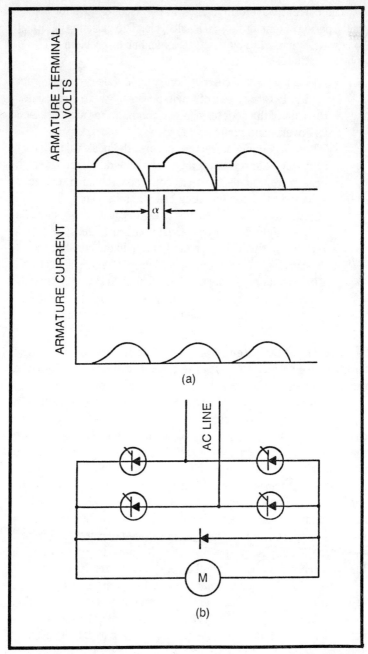

Fig. 9-17. At (a) voltage and current waveforms from a motor speed control system, and (b) a single phase, full-wave bridge circuit with a freewheeling diode.

The current waveform is seen to be discontinuous for a substantial period of time during each half cycle. The pulsating nature of the current waveform gives rise to a number of problems. These problems include:

- Form factor of the current is very high. Values ranging from 1.35 to 1.5 are typical of single-phase full-wave controllers.
- Because of the poor form factor, motor losses are increased and overheating may occur.
- When discontinuous current occurs, the peak value of current may be three times as great as the dc value. The high peak current gives rise to a proportionately high reactance voltage in the commutating coil when it occurs. The high reactance voltage, in turn, is likely to cause brush sparking with possible deleterious effects on brush life.
- Because of the high harmonic content in the current, the IR drop in the motor becomes an IZ drop. Speed is likely to fall with load more rapidly than would be the case with a steady value of current. The IR compensation in the controller must be increased to a point where oscillations may occur at low speed.

Because of the potential problems listed above it is necessary for the motor designer to recognize the implications of single-phase rectified power and to consider them during the design process.

The effort in design of the motor is to improve the current form factor. This is done by increasing the inductance of the armature circuit. For example, a doubling of armature inductance will reduce the current peak by almost one half. Motors designed for rectified power application usually seek to maximize the total armature circuit inductance, but this must be done in a way that does not increase the energy storage associated with an individual coil side. Otherwise commutation reactance voltage will be adversely affected. A common design approach is to minimize the number of poles used around the air gap. This approach produces the desired effect of maximizing the inductance for a given number of armature conductors.

It is possible to additionally improve the form factor by the use of an external inductance. Such an arrangement is shown in Fig. 9-18. If a large enough coil inductance is used, the armature current can be made continuous and the peak value of current significantly reduced.

The negative side of using an external choke coil is that it does represent an additional cost. Usually an amount of copper and steel

212

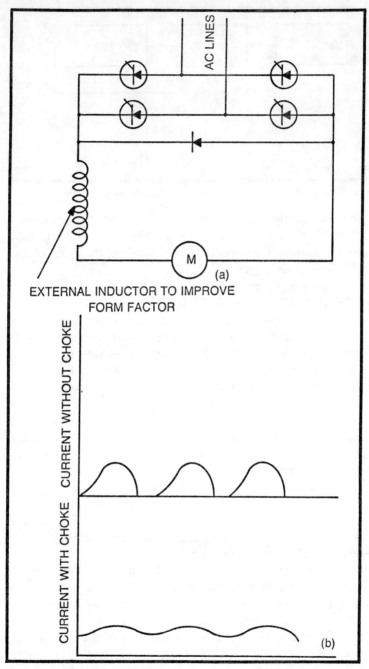

Fig. 9-18. At (a) bridge circuit with an external choke to improve form factor, and at (b) current waveforms with and without the choke in the circuit.

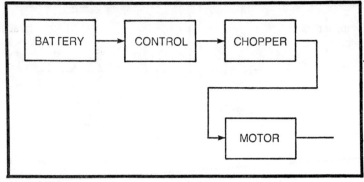

Fig. 9-19. Schematic of chopper control of battery powered motor.

equal to a fraction of that contained in the motor cores will make a choke coil adequate to produce the desired results. When a choke coil is required, its cost should be weighed against the additional cost of a three-phase supply.

DC TO DC CONVERSION, THE
STORAGE BATTERY AS A POWER SUPPLY

A large number of dc motors are used in mobile applications where the primary source of electric energy is a storage battery. In many of these applications it is necessary or desirable to be able to vary the speed of the motor. An example of such an application might be the traction motor in an electric vehicle. The driver of such a vehicle needs a means of controlling the motor speed. He must be able to accelerate the vehicle, to maintain a speed, or to stop it, as the situation might require. While a storage battery is a fixed voltage device, the requirements for speed control are such that a varying amount of the available battery voltage be applied to the motor.

While the shunt-field and PM motors are the most popular choices for use with rectified ac power, the series-field motor is popular for many battery powered applications. The series-field motor has the highest starting torque and accelerating capability of all and works well with the available controllers.

The variable resistance method of dropping the armature voltage of field current is one way of controlling motor speed, but new power electronic devices offer more attractive alternatives to inserted resistance techniques.

The problem of speed control involves taking electric energy from the battery at the fixed battery voltage level. The electric energy then converts it to a lower level of voltage which is adjusted according to the speed that is desired.

The power conversion is achieved by means of a switching amplifier (or chopper). The switching amplifier is turned on and off at a high repetition rate to alternately conduct and block current flow from the battery to the motor. The schematic of a simple chopper control is shown in Fig. 9-19. The effect of interrupting the voltage applied to the motor is to limit the amount of current that flows. The exact magnitude of the dc current is determined by the applied voltage, the winding impedance, and the switching rate.

Two main techniques are used with switching amplifiers. They are pulse width modulation (PWM) and pulse rate modulation (PRM).

PULSE WIDTH MODULATION CONTROL

In PWM, the chopper operates at a fixed frequency, alternately applying and then blocking the battery voltage to the load. The output of the chopper is a series of square pulses.

The flow of electric power to the load is regulated by controlling the width of the voltage pulse. As the load increases, the on time for current flow increases while the off time decreases proportionately. (Fig. 9-20). In this regard, PWM can be seen to have a similarity to phase control of rectified ac power.

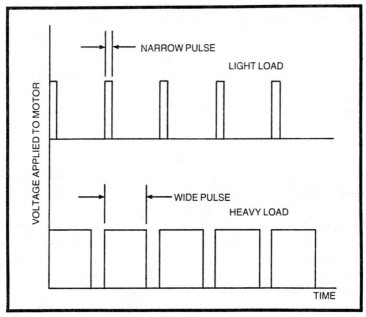

Fig. 9-20. Voltage wave forms with pulse width modulation.

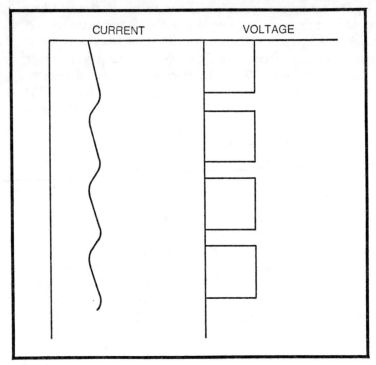

Fig. 9-21. Typical voltage/current waveforms from pulse width modulation chopper controller.

The switching frequency of a PWM chopper is usually between 1 and 2 kilohertz. The switching frequency means a period between pulses that is usually small compared to the electrical time constant of the motor. As a result the armature current is greatly smoothed and does not show discontinuity. Figure 9-21 shows a typical voltage and current waveform.

The switching element in a PWM chopper can be either a power transistor or a thyristor (SCR). The transistor is easier to work with due to its turn-off capability. The transistor, though, is more expensive in large current ratings. By way of contrast, the SCR does not have the capability by itself to interrupt current flow. It therefore needs an auxiliary commutation circuit to create a reverse bias voltage to turn it off. This commutation circuit usually includes an inductor of substantial size and cost. The SCR is, however, the more economical device at the larger current ratings.

The economics are such that transistor devices are most commonly used with drives requiring low amperage. The SCR becomes a viable choice when current requirements reach several hundred amperes.

The application of dc motors with chopper controllers is usually less stringent than with rectified power. Usually the battery power comes in relatively low voltage levels (48 volts or less) which reduce the motor commutation demands. In addition, the relatively smooth current waveform eliminates the current peaks which also would aggravate commutation with rectified power.

PRM CONTROL

Pulse rate modulation differs from PWM in that the square wave output has a fixed duration but the rate of pulse generation is controlled. As is the case with the PWM system, a signal proportional to motor speed is compared to the input signal. However, the control circuit then takes the error signal and provides an output whose frequency is proportional to the error. The pulsed output of a PRM chopper is shown in Fig. 9-22.

As in the case of PWM, either transistors or SCRs can be used as the switching element in a PRM. The same considerations as to the advantages of either type also apply.

The PRM system provides a voltage of widely variable frequency. The wide range of frequency makes it less adaptable for

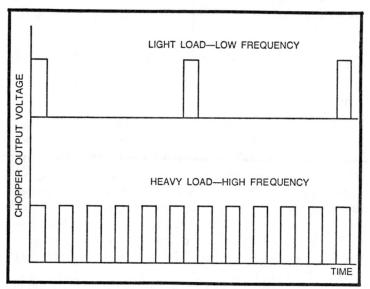

Fig. 9-22. Voltage wave forms with pulse-rate modulation.

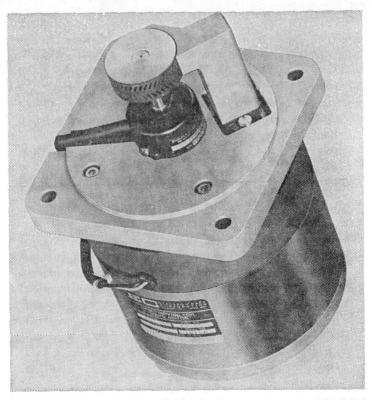

Fig. 9-23. A high performance servomotor used to drive the capstan of a digital tape drive system. (Courtesy of Electro-Craft Corp.)

motor control, where many motor characteristics are frequency sensitive.

SERVOMOTOR APPLICATION

The dc motor is an ideal choice as a servomotor because of the damping inherent in the negative slope of its torque-speed characteristic. This is especially true of the shunt-field and PM motors. As a result of this excellent compatibility, most high performance servo mechanisms have gone to direct drive dc systems. The performance levels of such servos can be truly amazing.

Figure 9-23 shows a motor with such outstanding characteristics. The motor pictured is used to turn and control the position of the capstan of a digital tape drive. The tape drive is required to read or record data using magnetic tapes or storage. The capstan motor may be required to make several hundred run cycles per second.

218

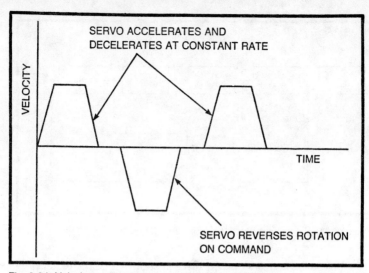

Fig. 9-24. Velocity profile of a servoed tape drive system.

Each operation requires the capstan to move the tape over the read/write head at a closely controlled speed to avoid distortion of the data transmitted. A typical velocity profile is shown in Fig. 9-24.

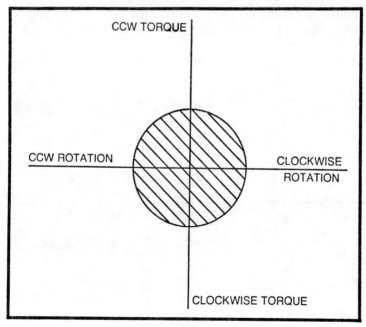

Fig. 9-25. Four-quadrant servo control.

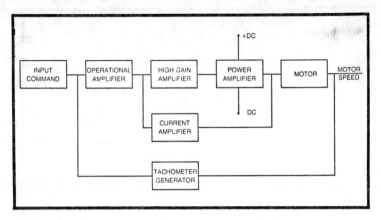

Fig. 9-26. The circuit schematic for four-quadrant control.

Figure 9-24 shows that the drive system is required to decelerate at a controlled rate. It must also have the capability of reversing rotation upon command. These are the characteristics of a true servo system and go by the name of four-quadrant control. A pictorial representation of four-quadrant control is shown in Fig. 9-25. Figure 9-25 can be contrasted with Fig. 9-15 to gain full appreciation of the added power of four-quadrant control. Such a drive can be programmed to perform very complex operations and

Fig. 9-27. DC motor with tachometer generator that couples directly to a rear extension of the motor shaft for a velocity feedback signal. (Courtesy Gould Inc., Electric Motor Division.)

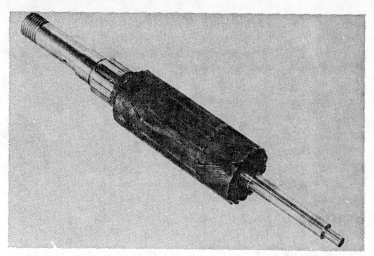

Fig. 9-28. Ironless construction of a low-inertia servomotor rotor. (Courtesy of Electro-Craft Corp.)

as a consequence is extensively used in computer peripheral equipment, machine tools, and automatic processing equipment.

The essentials of a four-quadrant control system are shown in Fig. 9-26. The heart of the system is the operational amplifier, which can reverse the polarity of the voltage applied to the motor. Very high performance servo systems require the use of a tachometer generator with only a negligible ripple in its output voltage. The best control is obtained by use of an optical tachometer that puts the speed sensing information into a digital form at a fixed voltage suitable to feed into control logic Fig. 9-27.

The best servo motors are built with permanent-magnet fields. The PM motor has the advantage of always having the field on, ready to go, even in the quiescent state.

In dc servo systems the bandwidth is limited by the inertia of the motor rotor. For this reason extreme design measures are resorted to in order to minimize inertia. An example of a rotor built without an iron core is shown in Fig. 9-28. For the same reason commutators, shafts, and bearings are designed with the smallest possible diameters. Motors designed with such criteria are not physically robust and must be handled carefully to avoid possible damage or impairment of performance.

Chapter 10

Testing, Maintenance, and Troubleshooting of DC Machines

The performance characteristics of dc machines are described by a set of well defined parameters. It is usual for the user of a machine to incorporate the performance characteristics he requires into a set of motor specifications. The specifications will define electrical as well as mechanical requirements.

Implicit in the specification of performance limits is the need to test the machine and measure the specific parameters. When testing is done it is important that standard procedures be used so that correlation can be made to test results conducted in another place and at another time.

This chapter describes how test values of the common motor parameters are obtained. It also describes good maintenance practices that will enhance good life and reliable service of the machine. Finally, it also discusses techniques useful in finding and repairing faults when breakdowns do occur.

ARMATURE RESISTANCE MEASUREMENT

The resistance of the armature circuit is an important factor in determining the amount of power loss and the resulting heating effect. It also has a primary importance in establishing the slope of the torque-speed curve as well as the electrical time constant.

Because of its importance to motor operation, armature resistance will usually be a production test on all motors.

In order to make a good test of armature resistance it is necessary to make the measurement directly from the commutator segments. Attempting to measure armature resistance through the brushes will invariably give results that are inconsistent. This is because of the relatively high resistance of the film between brushes and commutator.

The resistance of the armature may be measured from the commutator surface by a suitable meter, depending upon the range of resistance expected. Motors, down to small integral sizes, which are used in battery powered applications, will have very low resistance values and require a Kelvin Bridge for adequate sensitivity and accuracy of measurement. Motors smaller than one horsepower will have a resistance that can be measured accurately with a Wheatstone bridge instrument.

When making a measurement of resistance it is well to remember that it varies as a function of temperature. For this reason it is usual to specify resistance values at a standard temperature condition, such as 25°C (the average room temperature condition). If the test condition varies too greatly from the standard, it may be necessary to make a correction for it.

If armature resistance is to be measured it is best to perform the test before any other type of testing is done. If a motor has run for a period of time that allows it to reach a temperature stabilized condition, it may require a long cool down period to return the windings to room temperature. A large machine has sufficient heat storage in it so as to require several hours to cool off.

When measuring armature resistance it is necessary to place the probes of the measuring instrument on the proper commutator segments in order to obtain the correct results. To accomplish this the following points should be kept in mind (See Fig. 10-1):

- A wave connected armature has only two current paths through the armature regardless of the number of poles. This means that the circuit resistance can be measured with just two probes.
- If the armature winding is wave connected, the resistance can be measured by probes located 360/P degrees apart. Thus, for a four pole, wave connected armature, the probes would be located on commutator segments that are displaced by 90°.

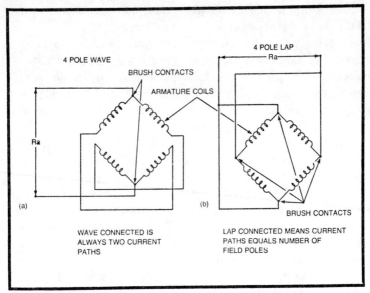

Fig. 10-1. The current paths through the armature, (a) wave connected and (b) lap connected.

- A lap connected armature has as many current paths through it as it has poles. It will therefore require as many probes to measure the resistance as there are poles.
- The proper location of the instrument probes to the commutator is at intervals equal to 360/P degrees. Alternate probes are of the same polarity and connected together.

ARMATURE INDUCTANCE MEASUREMENT

It is possible to make a direct measurement of armature inductance using an impedance bridge applied directly to the appropriate commutator segments as was done in making resistance measurements. Measurements taken with an impedance bridge tend to be very high and not representative of the actual circuit reactive power under rated conditions. The reason is that inductance is strongly influenced by magnetic circuit saturation and under loaded conditions, portions of the armature core do become saturated.

A more meaningful measurement of armature inductance is obtained by using an oscilloscope to study the current response to a step voltage input. This technique is actually a measurement of the electrical time constant of the armature at rated current condition. Accurate knowledge of armature resistance is also required. After measuring the time constant, the inductance is then calculated from the expression,

$$L = T_e R_a \qquad (10.1)$$

where

T_e is the electrical time constant
determined from the oscilloscope trace.
R_a is the armature resistance determined
by the method just described in the Armature
Resistance Measurement Section.

To make the measurement, the test circuit is arranged as in Fig. 10-2. The power supply voltage is adjusted to a value that produces full-load current in the armature. The resistance R_1 is of a value that is no greater than 1/10 that of the armature resistance. The measurement probes must make a good electrical connection directly to the commutator. The armature should be mounted in its normal position in the motor so as to produce the proper magnetic field effects.

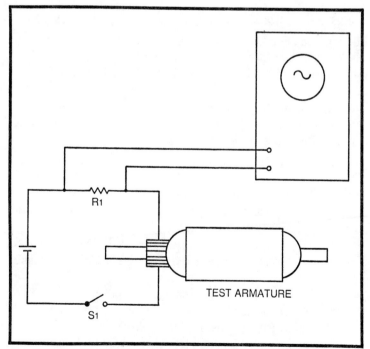

Fig. 10-2. Test circuit for the measurement of armature inductance.

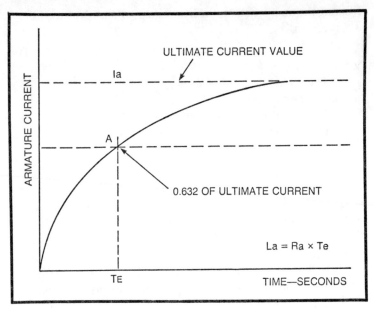

Fig. 10-3. Determination of armature circuit electrical time constant.

The oscilloscope leads will be connected across resistor R_1. The trace on the oscilloscope screen will be of the resistor voltage, which is, in turn, proportional to the current flowing through it. The oscilloscope is adjusted so that it is triggered by the closing of switch S_1.

A typical oscillogram trace will look like Fig. 10-3. The electrical time constant of the circuit is then determined in the following manner. (Referring to Fig. 10-3):

1. The ultimate value of the current is observed from the trace as in I_a.

2. The point on the vertical axis corresponding to $0.632 \times I_a$ is marked and a line drawn through it parallel to the horizontal axis.

3. The point of intersection, A, between the parallel line and the current response curve provides the electrical time constant.

4. The vertical line drawn from the point of intersection, A, to the horizontal axis allows the time constant to be read directly from the oscillogram scale.

5. The armature inductance s then calculate by using Eq. 10.1.

226

When making inductance measurements, it is necessary to make the proper connections between the power supply and the armature. The proper number of connections will vary depending upon the number of field poles and the type of armature connections used. The proper number of connections to the commutator is determined in the same way as for making the resistance measurements at the beginning of this chapter.

The armature must not be allowed to rotate while the oscillogram is being made. In the case of wound-field motors, rotation is prevented by connecting the power source directly to the armature and leaving the field coils unexcited. In the case of a permanent magnet motor, however, it is necessary to clamp the rotor shaft in order to prevent rotation when armature current flows.

TORQUE-SPEED PERFORMANCE

Motor speed and torque can be measured in a variety of ways. If the measurements are carefully made, it is possible to have good correlation between data taken by different methods. The equipment needed for testing the torque-speed characteristic must include a means of applying a measured load to the motor along with a speed measuring instrument.

The most commonly used device for loading a motor is a dynamometer. A dynamometer is nothing more than a dc generator mounted in a set of bearings so that it is free to rotate. The rotation of the dynamometer is restricted by the force measuring instrument, which can be a spring scale or load cell Fig. 10-4.

The dynamometer functions in the following manner. The armature of the dynamometer is connected to a set of load resistors. As the test motor drives the dynamometer at some speed, a voltage is generated that is proportional to the speed and/or field excitation. Current flow through the dynamometer's armature develops a torque which opposes the rotation of the test motor. The reaction force felt by the stator of the dynamometer is measured by the spring scale or load cell. The force measurement multiplied by the lever arm of the dynamometer gives the torque being transmitted from the motor.

The load applied to the test motor is determined by the armature current of the dymamometer. It can be adjusted by control of field excitation or the resistance of the load resistors. The output of the motor is dissipated as heat in the dynamometer and the load resistors. Care must always be taken to not overload the motor

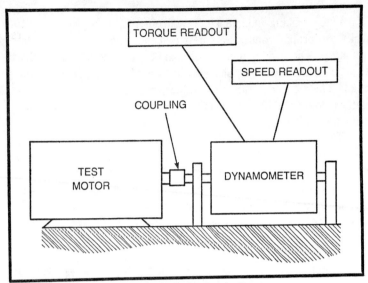

Fig. 10-4. The dynamometer method of torque testing.

being tested. Most motors have some overload capacity, but the duration must be limited to one minute or less to avoid dangerous overheating.

Dynamometers are used to test all sizes of motors, ranging from fractional horsepower sizes up to thousands of horsepower. They provide a convenient means of obtaining a continuous load adjustment. However, small motor sizes are often tested with nothing more than a pulley, a string, and a spring scale.

The spring and pulley method is illustrated in Fig. 10-5. A pulley of precisely known radius is affixed to the shaft of the test motor. A length of string is looped several times around the pulley. One end is attached to the scale and one end is held by a test technician. Load is appied to the test motor by pulling the string and developing tension in it. The increased string tension creates a friction with the pulley creating a load on the motor. The magnitude of this frictional load is read on the spring scale. The motor torque is found by multiplying the force reading of the scale by the radius of the pulley.

The spring and pulley method can be used with motors up to one horsepower. It is impractical with larger motors because of the heat dissipation between the pulley and string. The entire output of the motor is converted to heat at the interface of pulley and string.

Accurate torque data may be obtained with a variety of instruments. The most important factor in obtaining correlation of test data is the temperature condition of the motor. It must be kept in mind that torque and speed are temperature sensitive parameters (Fig. 10-6). In general, as motor temperature increases, the speed will also increase. On the other hand, the torque developed per unit of input current will decrease. For this reason it is necessary to test at a specified temperature condition.

A motor's rated torque and speed are based upon a stabilized temperature condition. But if a motor requires several hours to reach thermal equilibrium, the time may not be available to test at that condition. In such a case it is necessary to monitor the temperature of some portion of the motor structure. Torque can then be specified for a particular temperature condition.

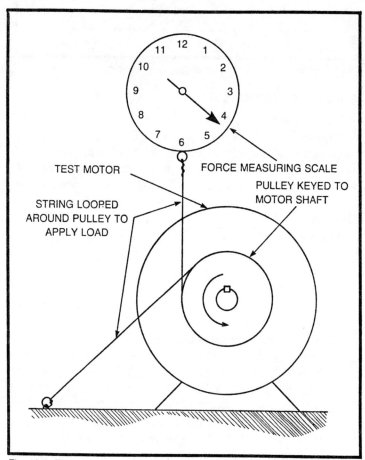

Fig. 10-5. The spring and pulley method of motor torque testing.

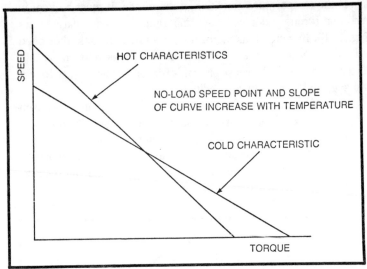

Fig. 10-6. Temperature sensitivity of torque-speed characteristics.

TESTING FOR ROTATIONAL LOSSES

It is often desirable to measure the rotational losses of a motor. These losses include windage and bearing friction, brush friction, hysteresis and eddy current losses in the core material, and damping losses due to circulating currents in commutating coils. If a thermal model is needed for purpose of analytical study it may also be necessary to segregate the components of rotational loss and afix a value to each.

A complete breakdown of rotational losses may be obtained by the following procedure.

1. A calibrated motor is used to drive the test motor. (A calibrated motor is one which has had its own torque-speed characteristics measured so that its output is accurately known.

2. The calibrated motor is used to drive the test motor at its rated speed. The brushes of the test motor are lifted or removed, and the motor is unexcited. The output power of the caibrated motor for this test condition is equal to the windage and bearing friction loss in the test motor. The losses are expressed in watts and are indicated by P_F.

3. The field winding of the motor is energized to rated current level. (If the motor has a series-connected field winding, it must be separately excited, with the vol-

tage adjusted to produce full-load current.) The calibrated motor is again used to drive the test motor at its rated speed. The output of the calibrated motor at this condition is equal to the windage and friction loss plus the iron loss (P_I) of the cores. Iron loss is obtained by subtracting P_F from the total loss obtained from this test.

A point to keep in mind is that a large component of the iron loss is due to eddy currents circulating in the core materials. The eddy currents are affected by the material resistivity. Since electrical steels have a large temperature coefficient of resistivity, the eddy currents are substantially reduced as the motor heats up. For this reason, the iron losses at full load and with stabilized temperature are likely to be considerably less than the value measured under no-load conditions.

4. Excitation is removed from the test motor and its brushes put into place at their specified pressure. The calibrated motor is again used to drive the test motor. At this condition the output of the calibrated motor is equal to the windage and bearing friction loss plus the brush friction loss, P_B. The brush friction is obtained by subtracting P_F from the total loss as determined by this test.

5. With the brushes in place, the fields are again excited to rated conditions as in procedure No. 3. The calibrated motor again drives the test motor at rated speed. The total output of the calibrated motor is equal to $P_F + P_I + P_B +$ (damping losses due to circulating currents, P_D). The damping loss is obtained by subtracting $(P_F + P_I + P_B)$ from the total loss.

The foregoing procedure allows a complete breakdown of the rotational motor losses. Since several of the loss components are temperature sensitive, temperature effect should be considered in applying the results to the other conditions.

GENERATED VOLTAGE TEST FOR TORQUE CONSTANT

This test procedure may be used in lieu of a dynamometer test to determine the torque-speed characteristics of permanent-magnet motors. It is necessary that the brushes be positioned on the neutral axis in order to obtain valid results.

In this procedure, the motor to be tested is driven in a generator mode by another motor. The test setup of Fig. 10-7 is used. The speed of the drive motor is adjusted to a convenient value

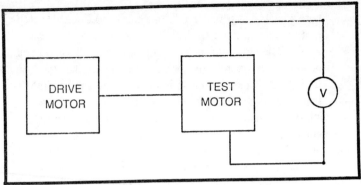

Fig. 10-7. Generated voltage method of testing torque-speed capability.

such as 1000 rpm. The generated voltage is measured at the output terminals of the test motor. The voltage constant of the motor is obtained by dividing the voltmeter reading by the drive speed.

$$K_E = \frac{V}{n}$$

Then the motor torque/amp capability is determined by the relationship

$$K_T = K \times K_E$$

where K is a constant whose value depends upon the units used for K_E and K_T (see Table 10-1). With the K_E and K_T values established by test measurement, a torque-speed curve can be drawn if the armature resistance (R_a) is known.

The no-load speed of the motor will be the rated voltage divided by the voltage constant.

$$N_o = \frac{V_T}{K_E}$$

Table 10-1. Constant Conversions.

Multiply K_E	by	To obtain K_T
volts/krpm	1.35	ounce-inches
volts/krpm	0.084	pound-inches
volts/rpm	7.04	pound-feet

The stalled torque is obtained by multiplying the torque constant by the ratio of terminal voltage to armature resistance.

$$T_S = K_T \frac{V}{R_a}$$

Since the torque-speed characteristic of a PM motor is linear, a straight line drawn between the two end points gives the desired performance data. The results of such a plot are shown in Fig. 10-8.

THE BLACK-BAND TEST FOR COMMUTATION

Black-band testing is a valuable test procedure that provides a quantitative measure of a motors capacity to commutate well under overload conditions. The procedure is applicable to machines equipped with commutating poles.

The test arrangement for making the test is indicated in Fig. 10-9. The procedure involves running the test motor at different levels of load and at each load point determining the level of commutating pole excitation at which brush sparking commences. The testing is done for both a "bucking" current and a "boosting" current. The current through the commutating field winding is adjusted by varying resistor, R_v. The current is made to "boost" or "buck" by reversing switch S_1.

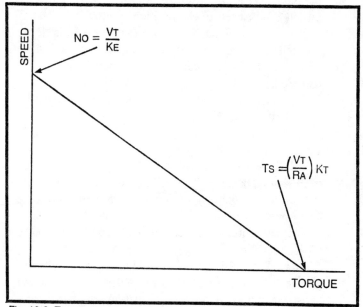

Fig. 10-8. Torque-speed curves obtained by two test methods.

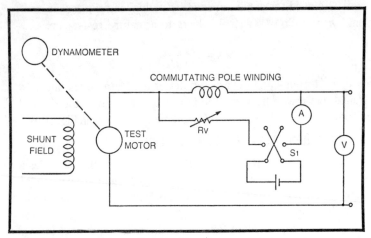

Fig. 10-9. Test setup for making black-band commutation test.

The procedure for a black-band test is as follows:

1. The test motor is arranged as shown in Fig. 10-8. Switch S₁ is in open position and the resistor Rv is adjusted for maximum resistance.

2. With rated voltages applied to armature and field windings, the motor is allowed to run at no-load.

3. Switch S₁ is closed to the position to provide a "boost" current to the commutating pole. Resistance Rv is slowly reduced. When the current through A reaches a value where brush sparking begins, the ammeter readings are recorded.

4. The motor is then loaded to 25% of its full load and the procedure of step No. 3 is repeated.

5. The procedure is again repeated at 50%, 75%, 100%, 125%, and 150% of full load.

6. Switch S₁ is then reversed so that the commutating pole excitation is "bucked." The bucking current is then slowly increased until sparking begins at 150% of full-load.

7. The values of bucking current are measured at each load point as the motor is unloaded to 125%, 100%, 75%, 50%, 25%, and no-load.

8. After all the test data has been obtained, the results are plotted to form the curves of Fig. 10-10. The buck and boost currents are expressed as a percentage of the armature current, A.

The curves obtained by black-band testing and interpreted as follows.

1. The area between the two curves represents the conditions of commutating pole excitation at which sparkless (black) commutation exists. It is the black band for the test motor.

2. The center of the black band should be at the horizontal axis at the no-load condition. If it is not, it is an indication that the brush axis is not located on the neutral axis. If the motor is intended for bidirectional rotation, the brush axis should be properly adjusted.

3. If the band converges it is an indicaiton that the machine is at the limit of its commutation capacity. This point of convergence should be to the right of the 100% load point.

4. If the band turns upwards as the load increases, as shown in Fig. 10-11(a), it is an indication that the normal commutating field is too weak. The condition can be corrected by reducing the air gap at the commutating pole by shimming behind it.

5. If the band turns downwards as load increases, as shown in Fig. 10-11(b), it indicates that the normal commutating field is too strong. The field should then be weakened by increasing the air gap at the commutating pole.

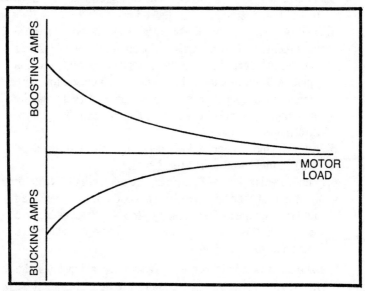

Fig. 10-10. Black-band test data plot.

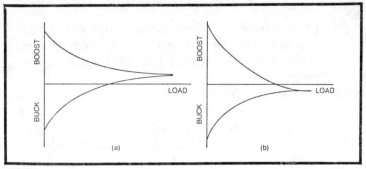

Fig. 10-11. Black-band test results, (a) too weak a commutating pole and (b) too strong a commutation pole.

DETERMINING THE DEMAGNETIZATION CURRENT FOR A PERMANENT-MAGNET MOTOR

It is often necessary or desirable to know the value of current at which a PM motor will start to lose magnetism. This procedure can be used to determine the ability of a PM motor to withstand high pulse currents.

1. The motor to be tested is arranged as in Fig. 10-12.
2. The voltage is adjusted so that the no-load speed of the motor is exactly 1000 rpm. The voltage is recorded.
3. The motor shaft is locked. The voltage is set so as to provide full-load current at locked rotor.
4. With switch S_1 open, the voltage is increased by a small increment. Switch S_1 is then closed and the oscilloscope used to measure the peak value of pulse current. Caution must be observed not to keep the switch closed for more than a few seconds since demagnetizing currents may be high enough to cause excessive, and damaging, temperature rise in the motor.
5. After the motor is pulsed with a current, the no-load speed is again measured as in step No. 2.
6. The alternate steps of pulsing with an increasing value of current and then retesting the no-load speed is repeated until the voltage as measured by step No. 2 has decreased by 5%. At this point the magnets may be considered as beginning to demagnetize.

Care should be taken in this test to avoid any heat buildup in the motor. The magnet materials are temperature sensitive and lose magnetic flux with increasing temperature. Thus, if there is significant heat build up it could cause an erroneous indication of demagnetization.

MEASURING ROTOR MOMENT OF INERTIA

It is often desirable to know the inertia of the motor's rotor. This is especially true for motors used in servo applications or in applications involving critical acceleration characteristics.

Rotor inertia may be measured by commercially available test instruments. These instruments, which are limited to small motor sizes of several pounds, provide a direct digital readout of the inertia measurement.

However, if an inertia measurment instrument is not available, the torsional pendulum method can be used to give accurate results. A stopwatch, a length of wire, and a known inertia standard are all that is required.

The torsional pendulum method works on the principal that its period of oscillation is proportional to the square root of its moment of inertia. Expressed mathematically, the expression becomes,

$$T = K\sqrt{J}$$

or

$$J = \left(\frac{T}{K}\right)^2 \tag{10.2}$$

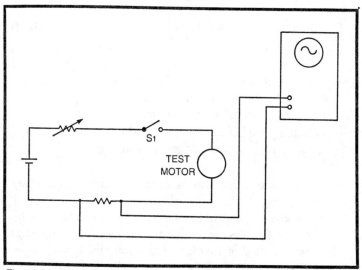

Fig. 10-12. Permanent-magnet demagnetization test arrangement.

237

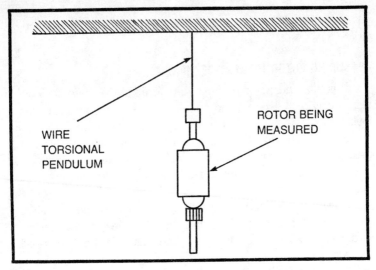

Fig. 10-13. The torsional pendulum method for measuring polar moment of inertia.

where J is the rotor moment of inertia
T is the period of oscillation
K is a proportionality constant

If a know inertia standard is available, its period of oscillation is determined. Then the period of oscillation for the test rotor is measured and the moment of inertia calculated by the expression,

$$J_M = J_S \left(\frac{T_M^2}{T_S} \right) \qquad (10.3)$$

where J_M is the moment of inertia of the test rotor
J_S is the known moment of inertia
T_M is the period of the test rotor.
T_S is the period of the standard inertia.

The detailed procedure works as follows:

1. A length of music wire is securely attached to a ceiling beam (Fig. 10-13).
2. The known standard is secured to the end of the wire by means of a collect or other clamping device. It is important that the axis of rotation of the standard and rotor be coincident with the wire.

3. The standard inertia is rotated about its axis by about 90° and then turned loose. It will then continue to oscillate about its axis. Care should be taken to not impart a swinging motion into the pendulum.

4. A fixed number of oscillations is timed with the stopwatch. At least ten cycles of oscillatory motion should be observed to make negligible any timing error. The time measured will be the T_S of Eq. 10.3.

5. The test rotor is then put at the end of the wire and steps 3 and 4 of the procedure are repeated. The same number of oscillations must be counted as was used in step 4. The measured time will be the value of T_M to be used in Eq. 10.3.

6. Having determined T_S and T_M by test measurement, and knowing the value of J_S, Eq. 10.3 is used to calculate J_M.

MEASURING SHAFT RUNOUT

Most applications of electric motors require that the output shaft turn reasonably true. If the shaft does not turn true it may not couple properly with its intended load or it can cause undesirable vibration.

Lack of control in critical manufacturing processes can create severe runout problems. This will occur if a shaft has a slight bend or if shaft diameters are not made concentric.

Because of its critical nature, shaft runout is usually a production test and is made in the following manner.

1. The motor being tested is supported securely so that it cannot move during the test.

2. A dial indicator of suitable sensitivity is located adjacent to the shaft to be measured. (Fig. 10-14).

3. The nib of the indicator is placed against the motor shaft within 1/16 inch of its end.

4. The shaft is slowly turned and the total excursion of the indicator is noted.

5. The total indicator reading (T.I.R.) should not exceed the specified amount. Most production standards will be equal to or better than 0.002 inch T.I.R. per inch of shaft extension.

MEASURING MOUNTING FACE PERPENDICULARITY

Motors used in applications where face mounting is used will usually have a perpendicularity specified between the mounting face

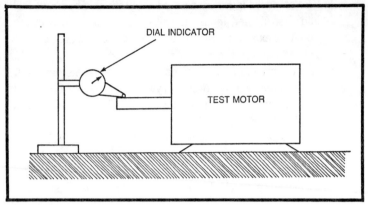

Fig. 10-14. Test setup for measuring shaft runout.

surface and the shaft axis of rotation. The perpendicularity is usually controlled by specifying a maximum runout of the mounting surface.

This characteristic is measured in the followiing manner:

1. A dial indicator of suitable senstivity is attached to the shaft of the test motor.
2. The indicator is adjusted so that its nib contacts the mounting surface (Fig. 10-15).
3. The motor shaft is slowly rotated and the total excursion of the indicator is noted.
4. The total indicator reading must not exceed the specified perpendicularity. A common standard is 0.002 inch T.I.R.

SEATING THE BRUSHES

When a new motor is assembled or when brushes are replaced in an old motor, in general there is a lack of intimate contact between the brushes and commutator surface. This will be the case even when the brushes are preformed to conform to the commutator radius. This condition will cause those areas which are in contact to carry an abnormally high current density under a full-load condition. The consequences of high current density can be very bad for brush and commutator. (See the Sparking from a Current Density Standpoint section of Chapter 8.)

For this reason it is important that new brushes be properly seated before they are subjected to full-load current. The easiest way to accomplish brush seating is by means of a commutator stone.

The commutator stone is a soft and crumbly material which only slightly polishes the commutator surface. It is used in the manner shown in Fig. 10-16. With the motor running at no-load the stone is pressed against the commutator surface on the side so the rotation is from the stone to the brush. As the stone crumbles against the commutator, particles are carried between the brush and commutator. While under the brush, the abrasiveness of the stone particles act to reduce the brush high spots and in the process create a complete seating.

When using a stone to seat brushes, the brushes should be removed periodically for inspection. The rate of seating will vary considerably from one motor to another depending upon the grade of brush used.

An alternative to stoning is sometimes used as a production process on small motors. After final assembly, the motors are simply placed on a run-rack and allowed to run at no-load for periods ranging from 8 to 24 hours. Under conditions of no-load, brush current density is very low and avoids the danger of excessive heat generation. The brush is then allowed to wear down mechanically until it becomes fully seated.

MOTOR MAINTENANCE

A dc machine is a rugged piece of equipment and if applied properly can be expected to give many years of service. Neverthe-

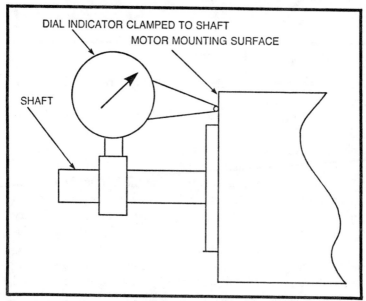

Fig. 10-15. Test setup for measuring mounting face perpendicularity.

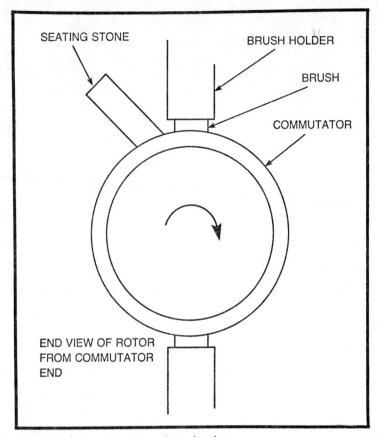

SEATING STONE

BRUSH HOLDER

BRUSH

COMMUTATOR

END VIEW OF ROTOR
FROM COMMUTATOR
END

Fig. 10-16. Using a stone to seat new brushes.

less, in the interest of reliable and trouble free service, some maintenance is required.

A regular motor maintenance schedule must be established on the basis of experience in a particular application. In general, maintenance can be keyed to the replacement of the carbon brushes.

The carbon or graphite brushes used in dc motors wear away at a fairly constant rate. It is important that the brushes be replaced before they have worn down so far that the brush shunt connection is pushed into the commutator. Brush life expectancy can and will vary considerably from one application to another. It is strongly influenced by environmental conditions such as temperature, humidity, and the presence of contaminants. A "normal" life might be many thousands of hours of running time or as low as several hundred hours, depending upon application conditions.

As the brushes are wearing down, it is normal for the commutator to also experience a similar but much slower wearing effect. The wear rate of the commutator will also vary considerably depending upon factors such as the abrasiveness of the brush and brush spring pressure. For a particular application there will usually be a consistent ratio between brush wear and commutator wear. This ratio can range from a value of 20:1 to as high as 100:1.

If the ratio of brush to commutator wear is at the lower end of the range, there is likely to be a noticeable groove in the commutator when the brushes are replaced. If this is the case, it is good practice to recondition the commutator surface. This is done by taking a light cut across the commutator while it is being turned. The turning/cutting operation will remove deficiencies in the commutator which have developed during operation.

If the ratio of brush-to-commutator wear is at the high end of the range, it is possible to go through several brush changes before commutator reconditioning is required.

When brush maintenance is being performed, it is also good practice to clean the inside of the motor. It is especially true in the case of enclosed motors that carbon dust will accumulate throughout the motor. Excessive buildup of carbon dust can cause problems by providing a low resistance path for electric current. A convenient method for cleaning dust out of the motor is to use an air hose to blow it out.

Regular maintenance procedures should also include a check of the motor's insulation system as a safety precaution. Checking the motor insulation resistance and dielectric breakdown may reveal a weakening of the insulation and help to avoid an electrical hazard.

MOTOR TROUBLESHOOTING

Following are a number of the more commonly encountered motor problems along with the most frequent causes. Very often the most obvious solutions to a problem are overlooked while more complex procedures are pursued.

Trouble
Excessive Motor Temperature:

Probable Cause

1. Motor is overloaded. Check load coupled to motor. Check bearings and gearing. Consider the effect of pulley ratio.
2. Motor is misapplied. Too many start and braking cycles will require a derated motor.
3. Ambient conditions. Consider heat buildup of adjacent equipment. Is ventilating air at elevated temperature.

4. Motor electrical fault. Shorted windings in field or armature circuits can cause overheating. Check winding resistances.
5. Motor bearing bad. Bearing can be tested by turning shaft by hand and feeling excessive drag.

Excessive brush sparking:
(Rapid brush wear)

1. Brushes not properly seated. Brushes should be fully seated by stoning or run-in.
3. Commutator surface poorly conditioned. Surface may be eccentric or individual bars may be high. Commutator should be turned and light cleanup cut made.
4. Wrong brush grade.
5. Commutating-pole strength is incorrect. Make black-band test.
6. Spring pressure is not constant.|Check brush holder for looseness. Check brush springs for proper pressure and secureness.
7. Contamination in environment. Consider presence of chemial contaminants that might react with commutator copper.
8. Shorted armature winding. Check armature resistance and make growler test.
9. Excessive mechanical vibration. Check dynamic balance of rotor.

Reversed Rotation:

1. Lead wires are improperly connected. Reconnect lead wires.

Excessive Noise and Vibration:

1. Lack of precision in load pulleys may be introducing vibration. Check runout of shafts and pulleys.
2. Rotor may not be properly dynamically balanced.
3. Motor mount may not be secure. Check mounting for secureness and rigidity.
4. Motor shaft may be misaligned with load. Alignment should be corrected.
5. Noise may be from noisy ball-bearings. Angular contact bearings should be pre-loaded with spring washer.
6. Bearings may be bad. Check bearings.

Excessive Motor Speed:

1. Main air gap may be too great. Steel shims should be placed behind magnetic poles to reduce speed.
2. Field windings may be shorted or opened. Check winding resistance against the specified value.
3. If PM motor, the magnets may be partially demagnetized from high armature current pulses. Condition can be confirmed by checking generated voltage and remagnetizing if necessary.

Motor Speed Too Low:

1. Motor is being overloaded. Check load.
2. Airgap is too small. Shims should be removed from behind the pole pieces.
3. Field winding may be opened or shorted.
4. Armature winding may be opened or shorted.

Glossary

alnico. A class of permanent-magnet material characterized by high flux density and relatively high cost.

armature. An assemblage of windings whose purpose is to interact with a magnetic field.

armature reaction. (1) The magnetizing (or demagnetizing) effects of the armature when excited by a current. (2) Angular acceleration of a rotor expressed in units of radians per squared second. (3) The magnetic flux density.

bidirectional control. A closed loop system capable of providing instantaneous correction with positive or negative torque.

black-band test. A test to determine a machines effectiveness in commutating overload currents.

brushes. The sliding contacts used to transfer electric current to and from the armature.

brush holder. The mechanical device used to support the brushes against the commutator surface.

ceramic magnet. A class of permanent-magnet material characterized by low flux density, high coercivity, and relatively low cost.

closed loop control (servo systems). A control system where the output is continuously monitored and feedback provided to the control amplifier to provide bidirectional control.

compound-field motor. A motor which has both a series-and shunt-field coil.

commutation. The act of switching current polarity as conductors move from one magnetic pole region to another.

commutator. A current collecting device whose function is to switch current polarity.

245

contact drop. The voltage drop that occurs between brushes and commutator, usually of several volts magnitude. Contact drop is a function of brush material, spring pressure, current density, and operating conditions.

controller. The control device used to adjust motor speed and provide constant speed from no-load to full-load condition.

dc chopper. A type of voltage regulator which controls the average voltage applied to a motor by rapidly opening and closing a solid state switch.

dc current. A unidirectional electric current that may be constant as with a battery supply or have a time varying characteristic as is the case with rectified ac power. The dc current is the average value.

demagnetization. A condition that can develop when armature reaction becomes stronger than the main field magnetizing strength.

E. The generated voltage of rotation.

electrical time constant. A figure of merit which expresses a circuits responsiveness to an applied step voltage. The value is expressed as the ratio of L/R.

electromagnet. An assembly composed of a coil wound around an iron core. When the coil conducts electric current, the core becomes magnetized.

end flux. A component of flux due to armature reaction which exists in the end-turn region of the armature. The end flux contributes to the commutation reactance voltage.

field. A term used to describe the magnetic condition of the air gap in an electric machine. Or, it is also used to describe the group of stator elements including the pole pieces, yoke, and field coils.

field coil. The electric winding in a motor which excites the main field electromagnet.

firing angle. The angle at which an SCR starts to conduct during the positive cycle of the ac line voltage.

form factor. An index used to describe the heating effects of time varying dc current. Form-factor is the ratio of rms current to dc current.

I_a . Armature current.

I^2R **losses.** Power losses due to the heating which occurs when electric current flows through a conductor of finite resistance.

inductance. A property of an electric winding which describes the amount of magnetic energy storage per unit current associated with it.

interpole (commutating pole). A special electromagnet assembly placed on the quadrature axis of a motor to assist the commutation function.

interpolar flux. A component of flux due to armature reaction which crosses the air gap along the brush axis and enters other members of the motor. It contributes to commutation reactance voltage.

J. The polar moment of inertia for a motor rotor usually expressed in terms of a torque multiplied by the square of a unit of time (example, $J = $ pound $\cdot$ feet $\times$ second^2).

K_E. Motor voltage constant.

K_T. Motor torque constant.

L_a. The inductance of the armature circuit.

magnetic field. A magnetic field is a region of space in which electrodynamic phenomena, such as electromagnetic force development or voltage generation, can occur. A magnetic field may be created by an electric current or a magnet.

magnetic flux. The amount of magnetism due to a source such as a circuit or magnet usually expressed as maxwells or lines.

magnetic flux density. The amount of magnetic flux in a given area expressed as gauss or lines per square inch.

magnetizing force. The strength of the magnetic source usually expressed in terms of ampere-turns per inch or oersteds.

main axis. The geometric axis of the motor along which the main field coils are symmetrically placed.

mechanical time constant. A figure of merit used to describe a motors accelerating capability. It is defined as the time needed to accelerate to 63.2% of the final stable speed.

N_o. Motor no-load speed.

neutral axis. The geometric motor axis half-way between the main field axes. It so named because motor will have equal bidirectional characteristics with the brushes on this axis.

no-load speed. The motor speed attained with full voltages applied and no load on the shaft.

overcommutation. The commutation condition when current is transfered too rapidly to the entering commutor bar. This condition may cause sparking at the leading edge of the brush.

P. The number of magnetic poles in a machine.

P_{em}. The developed electromagnetic power equal to the product of E and I.

permanent magnet. A material with properties that allow it to support an external magnetic field without electric current excitation.

PM motor. A motor in which the magnetic field is provided by permanent magnets instead of an electromagnet.

PRM (PFM). Pulse rate modulation (pulse frequency modulation) is a technique used to provide regulation with chopper controls.

PWM. Pulse width modulation is a technique used for regulation with a chopper control.

quadrature axis. See neutral axis.

R_a. Armature resistance.

rare-earth cobalt magnet. A class of permanent magnet material characterized by medium flux density, very high coercivity, and very high cost.

rectifier. A rectifier is a device that will conduct current in only one direction. When used in combination with an ac power supply it becomes an ac to dc converter.

regulation (R). The change of motor speed with changes in motor load usually expressed in units of torque per rpm.

ripple torque. The variation in the developed torque of a machine at constant speed as a percentage of the average torque. The ripple effect is caused by the time varying nature of discrete coil contributions which add up to produce the output torque.

rotational losses. The motor power losses associated with the turning of the rotor in the magnetic field, including rotor iron loss, windage and friction, and stray losses.

rotor. The rotating portion of a motor.

rms current. A measure applied to a time varying current to assess its heating effect. The rms value is the required dc current to produce an equivalent heating.

R_T. The equivalent analoque thermal resistance expressed in units of centrigrade degrees per watt.

SCR. A silicon controlled rectifier is a rectifier device which conducts current in the forward direction only upon receiving an electrical signal. Thereupon it continues to conduct until the current falls to zero.

series-field motor. A motor in which the field coils are connected in series with the armature so that the coil excitation varies with the armature current and motor load.

servomotor. A motor especially designed for use in a closed loop control system. The motor must have characteristics of low inertia and high acceleration in order to provide a high level of responsiveness.

shunt-field motor. A motor in which the field coil is connected directly across a supply voltage so as to provide a constant magnetic field strength.

slot flux. A component of flux due to armature reaction which crosses over the rotor slot without entering other motor members and contributes to the reactance voltage.

sparking. A generally undesirable condition which may occur under the brushes of a motor during commutation. Sparking is an indication of less than ideal commutation conditions.

stator. The stationary portion of a motor.

T,t. Motor torque.

tachometer generator. A dc machine used in a generator mode as a speed measuring device, in contrast to a power generator.

torque. The moment of force which tends to cause rotation in an electric motor. It is expressed in units of force and distance, i.e., pound-feet, newton-meter, etc.

undercommutation. The commutation condition when the current switching to the entering bar is delayed. This condition produces sparking at the trailing brush edge.

V_T. The voltage at the terminals of a machine.

voltage ripple. The variation in the generated voltage of an armature winding due to the summation effects of the discrete coil voltages. Ripple voltage is expressed as a percentage of the average generated voltage.

yoke. The yoke is the portion of a motor structure which connects the magnet poles and carries the lines of force.

Z. The total number of conductors in an armature winding.

η. Motor efficiency expresses the output power as a percentage of input power.

ϕ. Magnetic flux per pole usually in units of maxwells (lines).

Index